Fernando Grossklaus

# Onion cultivation

Fernando Grossklaus

# Onion cultivation

## Research, General Aspects of Cultivation, Soil Management and Nutrition

ScienciaScripts

Cover image: www.ingimage.com

This book is a translation from the original published under ISBN 978-620-2-17584-5.

Publisher:
Sciencia Scripts
is a trademark of
Dodo Books Indian Ocean Ltd. and OmniScriptum S.R.L publishing group

120 High Road, East Finchley, London, N2 9ED, United Kingdom
Str. Armeneasca 28/1, office 1, Chisinau MD-2012, Republic of Moldova, Europe
Printed at: see last page
**ISBN: 978-620-7-34055-2**

Table of contents:

# FEDERAL PUBLIC SERVICE
# MINISTRY OF EDUCATION

FEDERAL INSTITUTE OF EDUCATION
, SCIENCE AND TECHNOLOGY CATARINENSE
CAMPUS RIO DO SUL HIGHER
EDUCATION COURSE IN HORTICULTURE TECHNOLOGY

**Fernando Grossklaus**

# ONION CULTIVATION: RESEARCH, GENERAL ASPECTS OF CULTIVATION, SOIL MANAGEMENT AND MI'RICÁO

Supervised Internship Report carried out at EPAGRI - Ituporanga Experimental Station and presented to the Federal Institute of Education, Science and Technology - Rio do Sul Campus, as a partial requirement for obtaining the Certificate of Completion of the Higher Degree in Horticulture Technology, under the supervision of Prof. **Romano R. Valicheski** and Agr. Eng. Msc. **Claudinei Kurtz** with CREA/SC registration 033389-2

Rio do Sul-SC
2011

# SUMMARY

The aim of this work was to describe aspects related to onion cultivation through bibliographical research in books, magazines, articles and theses, as well as the implementation of experiments in the area of soil fertility, management and onion crop nutrition. During the 2011/2012 harvest, different experiments were set up at EPAGRI - Ituporanga-SC Experimental Station (EEITU) to evaluate the response to phosphate, potassium, organic and organo-mineral fertilization, with the aim of obtaining regional fertilizer recommendations and management. To do this, two experiments were carried out with increasing doses of Phosphorus (0, 50, 100, 200, 400 and 800 kg $ha^{-1}$ ), in soil with different levels of P; one of Potassium (0, 50, 100, 200, 400 and 800 kg $ha^{-1}$ ) and one experiment with organo-mineral fertilization (0, 4500, 9000, 18000 kg $ha^{-1}$ of poultry manure, mineral fertilization NPK 100, 120, 90 kg $ha^{-1}$ respectively, and poultry manure + mineral fertilization (NPK) in addition to organic fertilization). For the P experiment in the 2010/2011 harvest, in soil with a content of 4.0 mg/$dm^3$ , there was a significant response to fertilization in the commercial production of bulbs up to the maximum dose applied of 800 kg $ha^{-1}$ , with a commercial production of 40134 kg $ha^{-1}$ , with no influence on post-harvest losses. In the 2011/2012 harvest, in soil with 4.0mg/$dm^3$ of P (Mehlich), the responses behaved in the same way as the previous year, with an increase of 16099kg $ha^{-1}$ for the maximum dose applied, also influencing the percentage of class "2" bulbs with maximum totals for the treatment without fertilization, decreasing polymonally according to the dose applied. As for the soil, which had 26.5 mg/$dm^3$ of P (Mehlich), the doses applied had no influence on the commercial production of bulbs, the percentage of class "2" bulbs and the other variables evaluated. In the experiment where potassium doses were tested, the commercial production of bulbs was not influenced by the application of this element, nor were the other variables analyzed, following the same pattern in the 2010 and 2011 harvests, but it was observed in the field that the plots that did not receive potassium fertilization showed yellowing of the older leaves and drying of the tips. For the experiment on organic fertilization with poultry manure in the 2010/2011 harvest, the highest yield was obtained with 18,000 kg $ha^{-1}$ , an increase of 20.5% over the lowest dose used. When compared to mineral fertilization, organic fertilization obtained a 17.08% higher total yield. Organic fertilization plus mineral supplementation showed the same behaviour as mineral fertilization and did not differ from each other. As for the total loss of bulbs after storage, the treatment that received the highest dose of poultry manure showed the highest rate of loss (32.4%). For the 2011/2012 harvest, it was observed that the maximum yield was obtained with mineral fertilization with totals of 18924 kg $ha^{-1}$ (68.46%) higher than the control. Organo-mineral fertilization behaved in the same way, with 17915 kg $ha^{-1}$ (64.81%) more than the control. Organic fertilization with 18000 kg $ha^{-1}$ when compared to mineral and organomineral fertilization did not differ in commercial production and percentage of class "2" bulbs. The highest percentages of class "2" bulbs were observed for the treatment that received no fertilizer. Organomineral fertilization can be used not only to increase total bulb production, but also to improve the chemical, physical and biological conditions of the soil.

# INTRODUCTION

The cabbage onion (*Allium cepa L.*), a cultivated species belonging to the *Allium* genus and the *Alliaceae* family, is the third most produced vegetable in the world. In Brazil, the onion is also the third most important vegetable economically, with an average annual production of around 1,400 thousand tons of onion bulbs, a cultivated area of around 64,000 hectares and an average yield of 22,000 kilos per hectare. Production is concentrated in eight states, with Santa Catarina as the largest producing state, with average annual production of around 400,000 tons. It is an activity practiced mainly by small producers and more than 18,000 families have it as an economic activity on their properties.

Higher yields can be obtained with correct soil management, fertilization, pest, disease and weed control.

The amount of nutrients extracted by the onion crop varies according to productivity, cultivar, planting time, soil type, climate, geographical location and soil management. In general, the nutrient absorption curve follows the plant's growth pattern, with slow initial absorption, peak absorption at the full bulb stage and reduced amounts at the end of the cycle.

Fertilizer recommendations for onion crops are made at the state level or by Brazilian region. In order to verify the response of the onion crop to fertilization in a balanced way, different experiments were implemented in the 2011/2012 harvest at EPAGRI - Experimental Station of Ituporanga - SC (EEITU) to evaluate the response to phosphate, potassium, organic and organo-mineral fertilization, with the aim of obtaining regional fertilization and fertilizer management recommendations. To do this, two experiments were carried out with increasing doses of phosphorus (0, 50,100,200,400 and 800 kg $ha^{-1}$ ) in soil with different levels of P, one with potassium (0, 50,100,200,400 and 800 kg $ha^{-1}$ ) and also an organo-mineral experiment (0, 4500, 9000, 18000 kg $ha^{-1}$ of poultry manure, NPK Mineral Fertilization 100,120,90 kg $ha^{-1}$ respectively and Poultry Manure + Mineral Fertilization (NPK) in addition to organic fertilization).

The first part of this report covers: general aspects of onion cultivation related to origin, botanical description, climate, planting times, cultivar, cultivation systems, fertilizer recommendations, pest, disease and weed management, harvesting, storage and classification; and the second part: report of activities developed in the internship of the Higher Course of Technology in Horticulture at EPAGRI- EEITU in the period from July to December 2011, with the description and conduct of experiments carried out with fertilization and nutrition management in onion cultivation in the Alto Vale do Itajaí - SC, with results and discussion. This report is suitable for professionals in agricultural sciences and related areas, including agricultural technicians, technologists, agricultural engineers and onion producers.

# Chapter 1

## 1 HISTORY AND DESCRIPTION OF THE INTERNSHIP SITE

### 1.1 History

Linked to the State Government through the State Secretariat for Agriculture and Fisheries, the Santa Catarina Agricultural Research and Rural Extension Company (EPAGRI) came into being in 1991, when the Santa Catarina Agricultural Research Company S. A. (EMPASC), the Santa Catarina Credit and Rural Assistance Association (ACARPESC), the Santa Catarina Credit and Fisheries Assistance Association (ACARP) and the Santa Catarina Agricultural Research and Rural Extension Company (EPAGRI) were incorporated into a single institution.A. (EMPASC), the Santa Catarina Rural Credit and Assistance Association (ACARESC), the Santa Catarina Fishing Credit and Assistance Association (ACARPESC) and the Santa Catarina Beekeeping Institute (IASC) were incorporated into a single institution (EPAGRI, 2011).

Its administrative headquarters are located at Rodovia Admar Gonzaga, 1.347, Bairro Itacorubi, Florianópolis-SC. Consisting of twenty-three regional management offices, thirteen training centers, nine experimental stations and three specialized centers distributed throughout the state of Santa Catarina, among these units is the Ituporanga Experimental Station, located on Estrada Geral Lageado Águas Negras, which develops research and technologies available to the population. EPAGRI - Ituporanga Experimental Station (EEITU) was created on May 24, 1984, with the aim of participating in the processes of generating, adapting and disseminating agricultural technologies for the Alto Vale do Itajaí region, in conjunction with the rural extension and technical assistance team. It covers 32 municipalities with a predominance of small farms.

### 1.2 Site description

EPAGRI-EEITU has an area of 62 hectares for conducting field research, observation and demonstration units, producing seeds and seedlings, environmental preservation and reforestation. Its headquarters comprise a built-up area for the technical-administrative area and laboratories, and other areas to support research.

The personnel sector currently has 42 employees: 9 researchers; 1 chemist; 2 laboratory technicians; 2 agricultural technicians; 1 laboratory assistant; 6 administrative positions; 01 cleaning assistant and 20 farm workers.

He works mainly with: onions, various vegetables, forage crops, agroecology, sweet potatoes, forage crops, fruit growing, fish farming, soil management and plant nutrition, where most of the internship took place.

It has a soil analysis laboratory, a plant health laboratory, a weather station, an onion bulb warehouse, a cattle handling center and a multidisciplinary library.

# Chapter 2

## 2 REVIEW BIBLIOGRAPHIC REVIEW, REPORT AND DESCRIPTION OF ONION GROWING ACTIVITIES

### 2.1 Botanical Description and Center of Origin of the Onion (*Allium cepa L)*

The cabbage onion (*Allium cepa L.*) is a cultivated species belonging to the *Allium* genus and the *Alliaceae* family. This plant genus also includes other vegetable species of economic importance such as garlic (*A. sativum L.*), chives (A.*fistulosum L.)* and leeks [*A. ampeloprasum L. var. porrum*(L) J. Gay], as well as other ornamental species (BARBOSA, 2008) and (SOUZA & RESENDE, 2002).

The onion is a herbaceous plant with waxy leaves and hair-like roots (EMATER-RS, 1982). The commercial part is a tunicate bulb formed by the swelling of the leaf sheaths overlapping each other to form the reserve organ, made up of a shiny outer film and varying in color according to the cultivar (KASSAB, 1987).

Species belonging to the *Allium* genus are found in a wide range of altitudes and longitudes that extend from the Arctic Circle to the European continent, Asia, North America and Africa (BARBOSA, 2008). The exact center of origin of the *A. cepa* species is still controversial. The most likely center of origin is the arid regions of Central Asia, with poor, shallow soils subject to constant water stress (BARBOSA, 2008 ). According to Souza and Resende (2002), most botanists agree with the Russian geneticist Vavilov, who stated that Central Asia (now Iraq and Pakistan) was the primary center of origin. In Brazil, it was brought by European colonizers to the region of Rio Grande do Sul and from there spread to other states (RESENDE *et al.,* 2002).

### 2.2 Economic importance

Onions are the third most produced vegetable in the world. In 2009, according to FAO data, Brazil accounted for around 1.5 million tons, with a global production value of around US$ 317 million. Brazil and Argentina produce more than 60% of South America's onions and 97% of MERCOSUR's (OLIVEIRA, 2011).

In Brazil, onions are the third most economically important vegetable. Production in the 2009/10 harvest totaled 1,412.9 thousand tons of onion bulbs, with a cultivation area of around 64,167 hectares and an average yield of 22,090 kilos per hectare (EPAGRI, 2011). The planted area is concentrated in eight states: Santa Catarina, Rio Grande do Sul, Sao Paulo, Paraná, Bahia, Pernambuco, Minas Gerais and Goiás.

In the 2009/10 harvest, Santa Catarina had a gross production of 454.4 thousand tons, with a harvested area of 21,271 hectares and an average yield of 21,563 kilos per hectare, closing the harvest with only 298.1 thousand tons sold, due to a drop in yield and post-harvest losses (EPAGRI, 2011). It is an activity practiced mainly by small producers and its socio-economic importance is based not only on profitability, but also on the high demand for labor (EMBRAPA, 2011). In Santa Catarina, specifically in the Alto Vale do Itajaí region, the main producing municipalities are: Ituporanga (34.42%), Alfredo Wagner (14.67%), Aurora (8.29%), Imbuia (6.63), Petrolandia (5.67%), Chapadao do Lageado (4.08%), Vidal Ramos (3.53%), Leoberto Leal (3.44%), Atalanta (2.80%), Agrolandia (1.46%) totaling 85% of the state's production (EPAGRI, 2000).

In Brazil, the total number of families engaged in this activity is around 60,500 and in the state of Santa Catarina, more than 18,000 have it as their main activity (EPAGRI, 2000).

### 2.3 Climate requirements

Onions have a biennial cycle, with the vegetative phase giving rise to the bulb (1st year) and the reproductive phase (2nd year) giving rise to flowering and seed formation (SOUZA & RESENDE, 2002). For bulbification to occur, there must be an interaction between temperature and photoperiod (number of hours of light), with the photoperiod determining the limit of adaptation of the cultivars.

When a cultivar is exposed to a shorter photoperiod than required, no bulbs will form, but rather cigars (immature plants) will form even after long periods of growth (BARBOSA, 2008) and when the requirements are met, normal bulbs will form. On the other hand, plants exposed to long photoperiods and above the requirements of the cultivar will have early bulbification, i.e. if the plant has not had time to develop vegetatively then bulbs of reduced size and low commercial value will form (SOUZA & RESENDE, 2002).

According to the number of hours of light required for bulbification (adapted from BARBOSA, 2008), cultivars can be classified as follows:

1- Short-day (early) cultivars: require 11 to 12 hours/light/day;

2- Intermediate (medium) day cultivars: require 12 to 14 hours/light/day;

3- Long-day (late) cultivars: require more than 14 hours/light/day;

4- Very long-day cultivars: require more than 15 hours/light/day;

The photoperiod is considered the main factor in the induction, formation and maturation of bulbs, but its effects can be altered by temperature. Bulbification will only occur when the interactions between the two factors are satisfied. In other words, the length of the day for bulbification to begin decreases when there is an increase in temperature, but bulbification does not occur at high temperatures if the length of the day is insufficient (BARBOSA, 2008). Temperature can also influence nutrient absorption processes, conditioning the occurrence of biotic stress and the induction of flowering (COSTA *et al.*, 2002).

**Table 1:** Reaction of the onion crop to the effects of temperature and photoperiod.

| **Temperature** | **Photoperiod** | | |
|---|---|---|---|
| | Short | Normal | Long |
| Low | It doesn't bulb;<br>It can bloom; | It doesn't bulb;<br>It can bloom; | Delayed bulbification;<br>It can bloom; |
| Normal | It doesn't bulb;<br>It doesn't bloom; | Great production; | Reducing the cycle;<br>It forms smaller bulbs; |
| High | Does not bulb, does not flower | Reduction in the cycle; Smaller bulbs; | Large reduction in the cycle; Small bulbs; |

Source: (EPAGRI/CIRAM, 2011)

In addition to photoperiod and temperature, onions can also be affected by rainfall, humidity and nutrition. These can favor the speed of bulb development and affect the phytosanitary state and quality of the bulbs at harvest. Excessive humidity during the ripening and harvesting phase favors bulb rot, while dry weather contributes to bulb quality during harvesting and curing (BARBOSA, 2008).

### 2.4 Recommended cultivars for Santa Catarina

There are many cultivars used to grow onions in Santa Catarina, most of which were developed by EPAGRI - Experimental Station of Ituporanga - SC, covering early and medium cycle cultivars.

Early-cycle bulbs are sown in April/May and transplanted in June/July (depending on the location and altitude). Harvesting starts at the end of October and lasts until the end of November. They are notable for their shorter day length, softer tasting bulbs and they are not usually resistant to prolonged storage (BOEING, 2002). The Empasc 352 - Bola Precoce and Epagri 363 - Super Precoce cultivars stand out in this segment.

The medium cycle bulbs are sown in May/June and transplanted in August/September (depending on the location and altitude) and the harvest lasts from November until the first half of January. The bulbs form and ripen in longer days than the previous ones and have a spicy flavor, a more intense skin color and are well resistant to storage (BOEING, 2002). The cultivars that stand out are: Empasc 355 - Juporanga (medium cycle), Epagri 362 - Crioula Alto Vale (medium cycle).

The above cultivars stand out in southern Brazil and especially in the Alto Vale region due to their selection developed at EPAGRI - Ituporanga-SC Experimental Station, which seeks to develop materials adapted to the region's soil and climate characteristics.

In addition to the cultivars developed in Santa Catarina, other cultivars are also used for cultivation, as well as hybrids that have recently been tested and cultivated, especially those that adapt to the soil and climate conditions of the producing region, seeking high bulb yields and the quality required by the markets. More information on registered cultivars can be found at (MAPA, 2011).

### 2.5 Planting times

The planting time should be defined according to the physiological requirements of each cultivar, together with the local environmental conditions and the demands of the consumer market. It is determined by the photoperiod and the temperature required for the formation and production of quality bulbs (COSTA *et al.*, 2002).

Sowing for transplanting seedlings takes place from March to June. In Santa Catarina, sowing takes place from April until the end of June, with May being the most suitable month for sowing (SOUZA & RESENDE, 2002). The dates may vary according to the location of the growing area. Low-altitude regions have earlier sowing, followed by medium- and high-altitude regions. These differences are mainly due to the effect of temperature variations between regions.

### 2.6 Growing Systems

Onions can be grown using two main methods that are widespread in Santa Catarina: sowing in a seedbed followed by transplanting the seedlings using conventional tillage, minimum tillage and no-till, and direct sowing in a permanent location using conventional tillage or on straw.

#### 2.6.1 Transplanting seedlings

This method is more widespread in Brazil and Santa Catarina, mainly because it enables the production and selection of vigorous and healthy seedlings, which makes it possible to produce bulbs that are more uniform in shape and size (BARBOSA, 2008).

For the production of seedlings, beds of 0.15 to 0.20m in height and 1.0 to 1.2m in width and variable length are formed (Figure 2), with the application of chemical and organic fertilizers incorporated into the soil according to soil analysis. Sowing is done by hand or with the help of a seeder developed in the region (Figure 1) or by hand and covered with 1.0 to 1.5 cm of soil, sawdust, rice straw or rice husk ash. Sowing is done densely, using 2 to 3 grams of seeds/$m^2$ (BARBOSA, 2008). The cost of seeds for transplanting 1 hectare varies from 1.3 to 2.0 kg, considering 85% germination and 20% waste (SOUZA & RESENDE, 2002).

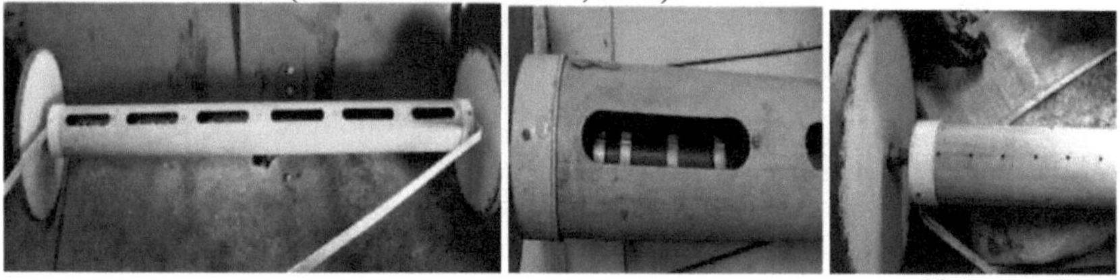

Photo: Grossklaus, 2011.

**Figure 1:** Manual seeder for onion seeds (left), seed compartment (center) and details of the holes through which the seeds come out (right).

To control pests and weeds, producers use a wide variety of pesticides to protect the seedlings, without worrying about the specificity of the product or dosage. More than 90% of producers use herbicides to control weeds, complemented by manual weeding (BOEING, 2002).

The seedlings are ready for transplanting 50 to 70 days after sowing, when they are 4-5 cm in diameter and 18 to 20 cm high (the size of a pencil) (Figure 2). The next stage begins with transplanting the seedlings, where there is a differentiation between the two production systems.

In the conventional system, most areas cultivated with onions still lack an integrated soil and water conservation system. Crop rotation is a practice that few do. Approximately 15% of producers use green manure to be incorporated when plowing the soil. Generally, the soil is extremely pulverized to facilitate transplanting the seedlings (Figure 4A), which is done manually (BOEING, 2002). Preparing the row for transplanting is basically done using a rotary hoe adapted with a kit for preparing two or four rows when using a micro tractor (Figure 2C), and five to seven rows when using a tire tractor. The rows are spaded 0.25 to 0.40m apart and then the seedlings are transplanted using 10-12 seedlings per linear meter.

Photo: Grossklaus, 2011.

**Figure 2:** (A) Seedling beds; (B) Seedling pad suitable for transplanting; (C) Details of the rotary hoe system attached to the microtractor;

Some onion producers in the municipality of Ituporanga-SC are testing an onion seedling transplanter (Figure 3). This machine is attached to a tractor and consists of 5 planting rows, where the seedlings are placed manually in a compartment made up of clamps on a rotating system that leads the seedling to the ground. It has a furrower that turns up the soil and then the seedling is introduced into the soil, ending with a wheel that accommodates the soil, allowing the seedlings to be transplanted. So far, according to some producers, the machine has some drawbacks that are making it difficult to use, such as: preparing and leveling the soil in the conventional system; lack of performance and efficiency; high cost and there has been no reduction in labor, which would be one of the conditions for opting to use the machine.

Photo: Grossklaus (2011)

**Figure 3:** Onion seedling transplanter (left) and details of the machine (center) and the transplant (right).

In the minimum tillage system, the soil is prepared in advance by planting cover crops (Figure 4B), the most commonly used being oats (*Avena spp*), mucuna (*Stilozobium spp)*, vetch (*Vicia spp),* sweetgrass or a corn + mucuna consortium. Days before transplanting, the green cover is desiccated using herbicides, and then tined with a knife-roller (BOEING, 2002). In recent years, there has been an increase in the use of millet (*Pennisetum glaucum*) grown alone or in a consortium with mucuna or sunflower (*Helianthus annuus*) as a cover crop, sown immediately after the onion harvest. When these plants reach physiological maturity, or the milky grain stage (sunflower), they are tilled with a knife roller or desiccated and left as mulch on the ground until planting. At EPAGRI - Ituporanga-SC Experimental Station, the use of rye (*Secale cereale* L.) as a cover crop is being tested. Its main characteristics are its precocity, hardiness, resistance to drought and low fertility soils (DERSPECH & CALEGARI, 1985 *apud* MONEGAT, 1991) and its high capacity to take advantage of less soluble elements (LEAL, 1970 *apud* MONEGAT, 1991). This extremely suitable plant is gaining ground and should be used by producers in the region.

Source: Grossklaus (2011)

**Figure 4:** Transplanting onions in the conventional tillage system (4A) and in the minimum tillage system, on straw from a cocktail of cover crops (4B).

### 2.6.2 Direct sowing

This method is used in Brazil mainly by producers with central pivot irrigation systems, located in the onion producing regions of the states of Minas Gerais, Sao Paulo, Goiás and Bahia (BARBOSA, 2002).

In the state of Santa Catarina, some onion growers in the main producing region have also invested in the introduction of high-precision direct seeding machines, which allow for a greater density of plants per unit area and, consequently, greater productivity and production per hectare cultivated (BOEING, 2002). In addition to the above, the lack of labor and labor complications have attracted growers to use this technique.

Sowing is done using precision machinery (Figure 5), in single or double rows, aiming for around 400,000 to 500,000 plants per hectare. This technique is limited to land with flat or slightly undulating topography, with medium-textured soils, free of spontaneous plants that are difficult to control, and the use of irrigation systems.

When areas cultivated with onions do not have the characteristics described above, there is a lack of uniformity in the emergence and *stand of* plants, which results in low yields. For this technique to be effective, it is necessary to plan at least one year in advance, so that soil analysis can be carried out and consequently liming and phosphating can be carried out. Weeds should be eliminated and cover crops used, which, as well as recycling nutrients, act as a biological herbicide and keep moisture in the soil, preventing erosion processes influenced by the mulch left on the ground. Details of this system can be seen below (Figure 5).

Source: Grosklaus (2011)

**Figure 5:** Details of direct seeding (A); Precision Vegetable Seeder (B and C);

## 2.7 Liming, Fertilization and Nutritional Disorders

When fertilizing and amending the soil, great care must be taken. In order to use fertilizers correctly and increase productivity, they should be applied in accordance with soil analysis data.

Soil sampling is considered the fundamental stage that precedes the actual analysis. It must be carried out in advance and certain precautions must be taken, such as: sampling times, depth, equipment, number of sub-samples and zig-zagging through the area, ensuring that the sample is as representative as possible of the area in question.

Sampling should be carried out at least 3 to 4 months before the crop is planted, taking into account the period in which the samples are analyzed and the return of the results, because if lime needs to be applied, it should be applied at least 2 months before the crop is planted.

### 2.7.1 Liming

When growing onions, liming is essential, due to the crop's high sensitivity to soil acidity. Liming should be carried out with the aim of neutralizing aluminium, as well as raising the pH to values close to 6.0, increasing the availability of nutrients.

Liming in the state of Santa Catarina is carried out according to the indications of two main methods: SMP index to reach pH 6.0 in water and base saturation.

The SMP index recommendation is based on the correlation between the SMP index and the potential acidity of the soil (H+Al). The lower the SMP index, the greater the amount of H+Al in the soil and the greater the amount of lime to be applied to achieve the right pH (TOMÉ Jr, 1997). The data from the analysis is then correlated with (Table 2) to obtain the amount of lime to be applied directly with a PRNT of 100%.

The base saturation method uses the following equation:

NC (t/ha) = CTC. (V2 - V1) / 100

In this way, we have NC, which is the need for lime to be applied at PRNT 100%; CTC is the cation exchange capacity; V2 is the percentage of the soil's base saturation that we want to achieve for crops, in the case of onions 80% (TOMÉ Jr, 1997); V1 is the percentage of the soil's base saturation, defined by the analysis. All the above data can be found in the soil analysis report (Annex

1). In general, liming is recommended when the base saturation is 10% below that recommended by the crop, and there is no need to apply it or the amount is very low, making it difficult to distribute it at higher values (TOMÉ Jr, 1997).

Liming should be carried out with uniform distribution over the whole area, either by hand or mechanically, and incorporated at a depth of 20 cm. Incorporation should be carried out with plowing followed by harrowing. Poor incorporation can lead to an imbalance between nutrients, resulting in variable pH ranges in the soil profile. A pH close to neutral or even alkaline can lead to a loss of $NH_4^+$ through volatilization, an imbalance in the K, Ca and Mg ratios and deficiencies or reduced availability (Figure 6) of micronutrients such as manganese, zinc, copper and iron (VIDIGAL *et al.*, 2002). In addition, the high concentration of limestone in the surface layers (0-10cm) neutralizes H and Al ions, raising the pH beyond the crop's needs. In addition to influencing nutrient absorption, poor incorporation can make it difficult for the roots to go deeper due to Al toxicity, which can harm the crop in the event of water shortages, as the soil loses moisture from the surface, thus compromising productivity.

**Table 2:** Liming recommendations based on the SMP index for correcting soil acidity in Rio Grande do Sul and Santa Catarina.

| **SMP Index** | **pH in water (1:1) to be reached** | | |
|---|---|---|---|
| | **5,5** | **6,0** | **6,5** |
| | **..........................................................ton/ha*** | | |
| 4,4 | 15,0 | 21,0 | 29,0 |
| 4,5 | 12,5 | 17,3 | 24,0 |
| 4,6 | 10,9 | 15,1 | 20,0 |
| 4,7 | 9,6 | 13,3 | 17,5 |
| 4,8 | 8,5 | 11,9 | 15,7 |
| 4,9 | 7,7 | 10,7 | 14,2 |
| 5,0 | 6,6 | 9,9 | 13,3 |
| 5,1 | 6,0 | 9,1 | 12,3 |
| 5,2 | 5,3 | 8,3 | 11,3 |
| 5,3 | 4,8 | 7,5 | 10,4 |
| 5,4 | 4,2 | 6,8 | 9,5 |
| 5,5 | 3,7 | 6,1 | 8,6 |
| 5,6 | 3,2 | 5,4 | 7,8 |
| 5,7 | 2,8 | 4,8 | 7,0 |
| 5,8 | 2,3 | 4,2 | 6,3 |
| 5,9 | 2,0 | 3,7 | 5,6 |
| 6,0 | 1,6 | 3,2 | 4,9 |
| 6,1 | 1,3 | 2,7 | 4,3 |
| 6,2 | 1,0 | 2,2 | 3,7 |
| 6,3 | 0,8 | 1,8 | 3,1 |
| 6,4 | 0,6 | 1,4 | 2,6 |
| 6,5 | 0,4 | 1,1 | 2,1 |
| 6,6 | 0,2 | 0,8 | 1,6 |
| 6,7 | 0,0 | 0,5 | 1,2 |
| 6,8 | 0,0 | 0,3 | 0,8 |
| 6,9 | 0,0 | 0,2 | 0,5 |
| 7,0 | 0,0 | 0,0 | 0,2 |
| 7,1 | 0 | 0 | 0 |

Source: Comissâo Química e de Fertilidade do Solo- RS/ SC (2004) * doses referring to a limestone with PRNT 100%

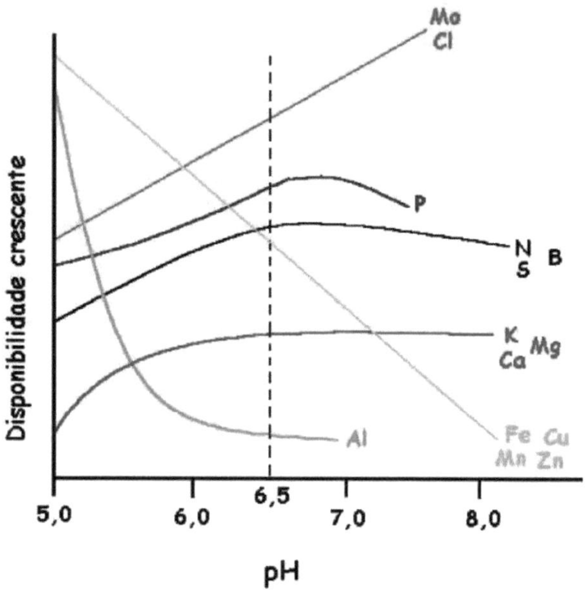


**Figure 6**: Relationship between soil pH and the availability of various nutrients. Adapted from Malavolta *et al.* (1989)

### 2.7.2 Fertilization

The supply of nutrients, in the right quantity, form and at the right time for the plant, is of great importance for most Brazilian soils, which have low natural fertility. In order to correct these problems and improve the supply of nutrients, it is necessary to know the chemical behavior of nutrients, which is influenced by the characteristics of the soil, environment and plant (VIDIGAL *et al.*, 2002).

Fertilizer recommendations for any crop should be based on plant demand, current or potential availability of the nutrient in the soil and the efficiency of nutrient recovery.

Fertilizer recommendations should always be made by interpreting the soil analysis, using the Fertilizer and Liming Manual for the states of Rio Grande do Sul and Santa Catarina. For phosphate and potassium fertilization recommendations, the data in Table 03 below is used:

**Table 3:** Phosphate and potassium fertilizer requirements for onion crops, used for interpreting soil analyses

| **Interpreted from the** | **Phosphorus** | **Potassium** |
|---|---|---|
| **P and K in the soil** | **Kg of P2O5/ha** | **Kg of K2O / ha** |
| Very low | 250 | 210 |
| Bass | 200 | 170 |

| | | |
|---|---|---|
| Medium | 160 | 130 |
| High | 120 | 90 |
| Very high | < 80 | < 60 |

In subsequent crops, apply 120 kg of P2O5 and 90 kg of K2O/ ha.
Source: CQFS-RS/SC (2004)

Phosphate fertilization can be carried out using formulated or simple fertilizers, always at the time of transplanting or sowing, following the previous recommendations in the planting line or applied by hand. Phosphate fertilizers should be applied and incorporated into the soil, due to their low mobility in the soil.

Potassium fertilization should follow the recommendations and can be applied together with phosphate fertilization when planting or sowing in the row or at the same time. When the recommended amount is more than 100 kg/ha of K2O, half should be applied at planting and the rest should be applied as a top dressing together with the nitrogen fertilizer. When the level of K in the soil is "high" or "very high", the recommended amount can only be applied as a top dressing together with the nitrogen fertilizer.

For nitrogen fertilization, you should follow the data in Table 04, which takes into account the level of organic matter in the soil.

| **Table 4:** Nitrogen fertilizer requirements according to soil organic matter content | |
|---|---|
| **Soil organic matter content %** | **Nitrogen Kg of N / ha** |
| < 2,5 | 95 |
| 2,6-50 | 75 |
| > 5,0 | < 55 |
| Source: CQFS-RS/SC (2004) | |

The nitrogen fertilizer can be applied in installments, with half the dose being used when the seedlings are transplanted and the rest at 45 days (CQFS-RS/SC, 2004). To improve the efficiency of the fertilizer and better use by the plants, dividing up the dose can be an alternative. According to KURTZ (2008), there was no interaction when the nitrogen dose was applied in installments as a base fertilizer followed by three top-dressing applications, but they did observe an increase in calcium and magnesium absorption and a lower number of crooked and broken leaves. When the dose to be applied is high, staggering the application becomes important, since a single top dressing at 45 days will lead to excessive vegetative growth to the detriment of the bulbs. In addition, in years of high rainfall, N may be lost through leaching and there may be a shortage at the time of bulbification, resulting in smaller bulbs (KURTZ, 2008).

More efficient N applications using the nutrient uptake pattern could be an alternative. Generally speaking, the amount of N to be applied will correspond to the amount of nutrient that the plant needs at each stage of the cycle, so as not to cause nutrient losses or yield losses due to lack or excess of N.

Numerous nitrogen fertilizers can be found on the market, in ammoniacal, nitric, encapsulated formulations, with the addition of micronutrients, as well as calcium and potassium. In view of the large number of fertilizers available, research was carried out at EPAGRI - Ituporanga-SC Experimental Station, in experiments carried out in 2010 (preliminary data) testing different sources of N available on the regional market, and no significant differences were found between the treatments.

### 2.7.3 Nutritional Disorders

Numerous nutritional deficiencies or toxicities can be observed in onion cultivation, mainly related to liming and the uncontrolled use of fertilizers. Details of the interaction and management of fertilizer and nutritional deficiencies are presented below.

#### 2.7.3.1 Zinc (Zn)

Plants deficient in Zn show reduced growth, striations on the leaves and also curving of the leaves with chlorosis (BARBOSA, 2008), usually manifested in the younger parts of the plants (MAGALHÂES, 1988). The conditions that predispose to deficiency are mainly related to excessive liming followed by an increase in pH and high concentrations of P in the soil solution (SOUZA & RESENDE, 2002). If a deficiency does occur, it can be corrected by applying Zinc Sulphate ($ZnSO_4$) at a dose of 15 to 20 kg $ha^{-1}$ via soil or foliar application at a dose of 0.5% of the volume of syrup. KURTZ

(2008) obtained a 14.5% increase in yield with the application of 2.0 kg $ha^{-1}$ of Zn added to the soil compared to the control, even in soil with a content considered sufficient for the crop.

#### 2.7.3.2 Boron (B)

It shows reduced growth, twisted, thick, brittle leaves and dry parts from the apex (BARBOSA, 2008; SOUZA & RESENDE, 2002). The root system is the first part of the plant to be affected by the deficiency, as well as growth points, differentiated areas and cell expansion organs (SOUZA & RESENDE, 2002; MAGALHÁES, 1988). Excessive liming, in soils with a low Organic Matter content (MAGALHÁES, 1988) and excessive potassium fertilization can lead to boron deficiency. In the event of a deficiency, use 15 to 20 kg $ha^{-1}$ of Borax or Boric Acid via the soil, since foliar fertilizations are less efficient than using the soil. KURTZ (2008) obtained no response to the application of this nutrient in soil with an organic matter content of 38 g $kg^{-1}$ .

#### 2.7.3.3 Manganese (Mn)

Symptoms commonly observed are internerval chlorosis, progressing from light green to yellow and light yellow mainly on the youngest leaves (SOUZA & RESENDE, 2002). Predisposing factors are peaty soils, calcareous mineral soils with a high pH and poor drainage (SOUZA & RESENDE, 2002). It is mainly influenced by excessive liming, causing the soil pH to rise above 6.5 (VIDIGAL et al., 2002). If a deficiency is observed, it can be corrected using manganese sulphate ($MnSO_4$) at 1% of the spray volume. KURTZ (2008) found no response to the application of this nutrient in soil with a pH of 6.0 and an average organic matter content.

#### 2.7.3.4 Copper (Cu)

Symptoms of deficiency are weak, stiff plants and leaves with whitish chlorosis, twisted and with necrosis at the tips (BARBOSA, 2008). Excessive liming with increased pH, high doses of P, and plants well nourished with nitrogen can reduce the availability of copper (VIDIGAL et al., 2002).

#### 2.7.3.5 Molybdenum (Mo)

An uncommon deficiency, the symptoms are leaves with marginal chlorosis or between veins, the first symptoms being similar to N deficiency. Older leaves are chlorotic or spotted, with veins remaining light green and necrotic leaf margins. The simple practice of liming, raising the pH above 5.5 in acidic soils, increases availability (SOUZA & RESENDE, 2002). Microbial activity can increase the Mo content in the soil solution made available by the displacement of adsorbed Mo into the solution, occupying spaces by organic compounds (BISSANI et al., 2008).

#### 2.7.3.6 Aluminum (Al)

The vast majority of plants are sensitive to aluminum toxicity, primarily because it damages the normal functioning of the roots, inhibiting their growth and blocking the mechanisms for acquiring and transporting water and nutrients (ROSSIELO & JACOB NETTO, 2006). The onion crop is relatively sensitive to soil acidity and aluminum toxicity. Among the main symptoms of toxidity are: reduced development, followed by thickening and few branching of the roots; darkening of the roots; low concentration of P in the plant, due to reduced absorption and translocation in the plant; reduced development of the aerial part of the plant caused by the factors described above (BISSANI et al., 2008). Controlling the pH by applying lime and raising the pH of the soil to 5.5 is enough to eliminate toxicity.

#### 2.7.3.7 Calcium (Ca)

The deficiency occurs mainly in the growing points of the area and the root. Deformation of new leaves, chlorosis and death of growth points. Factors predisposing to deficiency: rapid growth at high temperatures, low water content in the soil and antagonism with other cations such as $NH_4^+$ , $K^+$ and $Mg^{++}$ .

Photo: Grossklaus, 2011.

**Figure 7:** A and B - Zinc deficiency; C and D - Boron deficiency; E and F - Toxic effect of aluminum on onion roots.

## 2.8 Weeds

The development of onion plants is usually slow, especially in the early stages of growth, due to the reduced number of leaves, their cylindrical shape and erect habit, and the low size of the plants, which provide a low leaf area index and light interception during the crop cycle (BARBOSA, 2008). The crop irregularly covers the soil, allowing weeds to germinate at any stage of development (SOUZA & RESENDE, 2002).

The onion crop's ability to compete with weeds, whether transplanted or sown directly, is very low. The period before weeds interfered with the onion crop observed by SOARES et al., (2003) for transplanted onions was 42 days and living with weeds for the first 98 days reduced onion yields by 95% and average bulb weight by 91%. In direct sowing, weed competition is even more damaging, reducing yields by 30, 68 and 94% if competition after emergence lasts four, five and six weeks respectively (SOUZA & RESENDE, 2002).

The analysis of the period in which weed species compete with onion plants for production factors basically depends on the characteristics of the crop (cultivar, sowing or transplanting density) and the population and species of weeds present in the growing area (BARBOSA, 2008).

Several species of weeds are found growing onions in the Alto Vale region (Figure 8), some of which are difficult to control using physical, mechanical or chemical methods. The main plants that compete with onions and are difficult to control whether sown, transplanted or sown directly are: Tiririca (*Ciperus sp)*, wild garlic (*Nothoscordum inodorum*), wild celery (*Apium leptophyllum),* wild wormwood (*Artemisia Verlotorum* Lamotte), labarga (*Rumex sp),* chives (*Sisyrinchium laxum* Sims), winter pasture (*Poa Annua* L) and clover (*Oxalis sp)*.

Photo: Grossklaus (2011)

**Figure 8:** Weed species observed in the onion crop at the Ituporanga Experimental Station. A- *Oxalis sp* ; B- *Apium leptophyllum* ; C- *Artemisia Verlotorum* Lamotte ; D- *Poa Annua L*; E- *Ciperus sp*; F- *Sisyrinchium laxum* Sims

Weed control practices include weeding, hand picking and the use of herbicides. Due to the increase in plant population and the spread of direct sowing areas, weeding has become difficult and costly, leading producers to opt for the use of herbicides, which do not always effectively control all the species listed above. One of the major problems faced by farmers in the region is related to the control of dicotyledons.

The use of herbicides is frequent from the sowing stage until the bulbs are harvested. The main herbicides used by producers in the Alto Vale do Itajai-SC are listed in Table 05.

**Table 5:** Main MAPA-registered herbicides used to control weeds in the onion crop

| Product Commercial | Active ingredient | Dosage Kg or lt/ha | Controlled plants | Time of application |
|---|---|---|---|---|
| Herbadox 400 EC * | Pendimenthalin | 2,5 a 4,0 | Grasses and dicotyledons | Pre-emergency |
| Flumizin | Flumioxazine | 0,12 a 0,18 | Dicotyledons and grasses | Initial pre- and post-emergency |
| Totril * | loxynil | 0,5 a 1,0 | Dicotyledons | Post-emergency |
| Select * | Cletodim | 0,3 a 0,5 | Grasses | Post-emergency |

* Products used at the Ituporanga-SC Experimental Station;

## 2.9 Main pests

The onion crop can be attacked by a number of pests of economic importance. At the sowing stage, due to various factors, mainly climatic, there are no major pests. Pest attacks are mainly observed in the post-transplant phase, when temperatures rise and the number of pests occasionally increases. The pests of economic importance will be presented below.

### 2.9.1 *Thrips - Thrips tabaci*

*Thrips* (*Thrips tabaci Lind.*) or louse as it is known in the Alto Vale region is the main pest of onions and can cause losses of up to 50% of production according to SOUZA & RESENDE (2002). It is a small insect (1 mm long) in its adult stage (Figure 9). The nymphs are lighter in color, do not fly and shelter in the inner part of the leaves (sheath). They live in colonies and feed on leaf sap. The complete cycle lasts approximately 15 days, varying according to the temperature.

The attacked plant shows a silvery to whitish coloration, twisting of the leaves, yellowing and drying of the leaves from the apex to the base. Consequently, the plants do not fall over when they reach physiological maturity, which favors the accumulation of rainwater, causing rotting (GONQALVES, 2006). High infestations lead to a reduction in the size of onion bulbs.

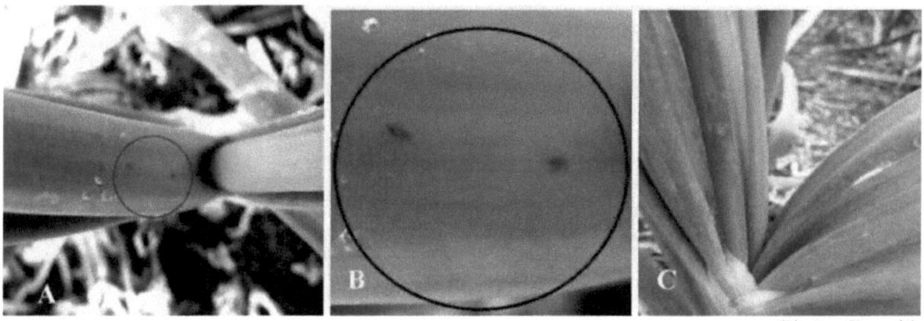


Photo: Grossklaus, 2011.

**Figure 9**: Presence of thrips attacking onion plants (A and B); Damage caused by the pest attack (C)

High population densities of lice in the Alto Vale do Itajaí-SC are observed from mid-October, with a peak at the end of November, varying according to the time of transplanting and the cultivar. The level of economic damage in Ituporanga-SC is 15 before and 30 after bulbing for the Criolla cultivar (GONQALVES, 2006).

It is a pest that is difficult to control due to its biological characteristics, its location on the plant, its leaf architecture, the waxy nature of the leaves, which makes phytosanitary treatment difficult, and it is favored by high temperatures and little rainfall, which leads to an increase in the insect's population.

Control should begin with soil management, seeking a nutritional balance with plants that are tolerant to the attack. The use of no-till farming, conservation practices and an increase in organic matter in the soil reduces the incidence of the pest. Exaggerated use of nitrogen fertilization can favour the attack due to its relationship with other nutrients (GONQALVES, 2006).

For the Alto Vale do Itajaí-SC region, control can be achieved by using early-cycle cultivars, avoiding consecutive plantings and adopting chemical control strategies when the damage level is reached.

Chemical control is the method most often used by producers in the region to control the pest, using the different active ingredients listed in Table 5.

Chemical control doesn't always completely control the pest, especially when the weather conditions are favorable to the pest and especially in cultivars with a medium or late cycle, in addition to the architecture of the plants which shelter the pests in their sheaths, making control difficult, spraying at night or in the late afternoon, when the temperature is lower and the pest comes out of the sheaths and is the target of spraying. The use of 27 fan spray tips, increased spray volume and rotation of active ingredients all contribute to more efficient control (SOUZA & RESENDE, 2002).

### 2.9.2 Threadworm caterpillar - *Agrotis ipsilon* (Hufnagen)

This pest is commonly found in the region's crops and does not suffer severe attacks. The caterpillars are cylindrical, grayish-brown in color, named after their habit of curling up when touched (Figure 10). They are nocturnal and take shelter in the soil during the day. Damage is characterized by the cutting of plants in the neck region (GONQALVES, 2006).

Control can be carried out by rolling the cover crops or preparing the soil before transplanting. In the field, chemical products are used (Table 6) when attacks are significant, by spraying at the end of the day with high volumes of spray.

Photo: Grossklaus (2011)

**Figure 10**: Damage caused by *Agrotis ipsilon* on the left and right: damaged plant and the presence of the pest.

### 2.9.3 Onion Fly - *Delia platura* (Meigen); and *Pseudosciara pedunculata* (Enderlein)

Onion crops can be attacked by the larvae of two different fly species.
The adults of *D. platura are* similar to houseflies, 8mm long, greyish-black in color and recognizable by their slow, low flight close to the ground surface. The larvae are yellowish white and 5mm long. They attack plants in the seedbed stage, when the plants have 2 to 3 leaves. The seedlings are perforated and turn yellow and can fall to the ground with total loss. In the field, they attack soon after transplanting, causing wilting, yellowing and rotting of the leaves.

This is more evident during periods of drought (GONÇALVES, 2006).

Factors that contribute to the attack are related to the decomposition of plant material from cover crops or the use of undecomposed organic material, which attracts the fly that feeds on undecomposed plant remains. They feed on these materials and can attack plants that are suffering from stress due to the planting conditions.

The *P. pedunculata* fly also has saprophytic habits. They have a thinner body, measure 8 to 9 mm in wingspan and 5 mm in length and are black in color. The larvae have a thinner body and are larger, measuring 8 to 9 mm in length. Damage is only seen after transplanting, unlike *D. platura, which* causes external tears in the root system (Figure 11) and does not penetrate the plants. Attacked plants show yellowing, browning of the central leaf and early bulbification. This damage is related to the stress suffered by the plants shortly after transplanting, causing partial death of the roots, which attracts the insect.

In recent years, there has been an increase in attacks by this pest in the Alto Vale do Itajaí-SC region, mainly due to a lack of crop rotation, the use of fresh manure and transplanting over cover crops that have recently been desiccated or incorporated into the soil.

It is difficult to control the pest once it has been attacked, as the larvae live in the soil. Strategies should be adopted for managing cover crops before transplanting, avoiding the use of undecomposed organic material, as well as avoiding the use of irrigation immediately after transplanting when there is no rain, thus favoring the setting of the seedlings and reducing stress and, consequently, the attack by the larvae. Adults can be controlled with the chemical products shown in Table 6.

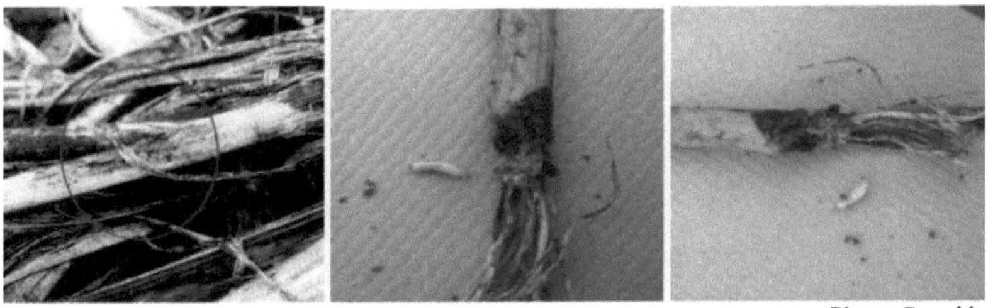

Photo: Grossklaus (2011)

**Figure 11:** Damage caused by onion fly larvae, observed in the onion crop, EPAGRI- Ituporanga-SC Experimental Station, 2011

**Table 6:** Main products used for pest control in onion cultivation

| Commercial Product | Active ingredient | Dosage (Kg or lt / Ha) | Controlled Pests |
|---|---|---|---|
| Karate Zeon 50 CS * | Lambda-cyhalothrin | 0,1 | Thrips |

| | | | |
|---|---|---|---|
| Lorsban 480 BR * (a) | Chlorpyrifos | 0,5 a 1,5 | Thrips, caterpillars and onion flies |
| Decis 25 EC | Deltamethrin | 0,25 | Thrips |

Source: Agrofit (2011)
* Products used at EPAGRI - Ituporanga-SC Experimental Station;
(a) Product not registered with MAPA for onion cultivation;

## 2.10 Main complaints

Numerous diseases attack the onion crop and can be caused by fungi, bacteria and viruses. The attack is even worse due mainly to the lack of resistant cultivars and climatic aspects in the Alto Vale production region. On a scale of economic importance, some diseases are considered to be significant: *Damping-off*; tip burn; mildew; purple blotch.

### 2.10.1 *Damping Off*

The toppling of seedlings during the initial germination phase is commonly found, causing damage mainly to the final plant stand. There are different causal agents and it is also favored by abiotic causes.

The causal agents of *damping-off* are *Pythium spp, Phytophora spp, Rhizoctonia solani, Fusarium oxysporum, Pyrenochaeta terrestris* and *Colletotrichum spp.* The initial symptoms are necrosis of the base of the plant tissues and toppling followed by rapid rotting, as can be seen in (Figure12).

Conditions of high soil humidity and low temperatures contribute to the attack, mainly because the fungi that cause toppling, with the exception of *Colletotrichum spp*, are inhabitants of the soil and are favored by the conditions described above. Other factors predispose to attack such as: waterlogged soils, excessive irrigation and poor quality water, shady locations, very dense and/or deep sowing, undecomposed organic matter, excessive covering material for the beds (SOUZA & RESENDE, 2002).

Source: Grossklaus (2011)

**Figure 12:** *Damping-off* in onion seedlings, observed on a property in the municipality of Aurora-SC.

In order to control the disease, the factors mentioned above must be controlled, seeds of guaranteed origin must be used, seeds must be treated and when this is not enough, fungicides in the form of Captan + Metalaxyl can be used to treat the soil before sowing (SOUZA & RESENDE, 2002).

### 2.10.2 Grey burn - *Botrytis squamosa* Walker

Tip blight, caused by *Botrytis sp,* is one of the most important diseases of the onion crop, causing serious damage mainly in the states of Rio Grande do Sul and Santa Catarina. The disease is favored by conditions of low temperature and high relative humidity, (SOUZA & RESENDE, 2002) a condition commonly observed in the Alto Vale do Itajaí-SC, in the sowing phase or initial phase in direct sowing, where there are heavy fogs followed by strong sunny spells. Relative humidity of 100% and a temperature of around 12 to 16° C for at least 24 hours are optimal factors for infection to occur, in addition to injuries caused by mechanical damage (wind and irrigation), insects and sunburn (SOUZA & RESENDE, 2002).

The symptoms are difficult to diagnose, as the main characteristic is dry tips, symptoms caused by water stress, excess moisture in the soil and ozone damage. Attacked plants show small white spots, numerous and followed by the death of the tip (Figure 13), and in flowerbeds it can cause the death of new plants (WORDELL FILHO & BOFF, 2006). They are characterized by small isolated spots on the leaf blade, with silvery halos and the rest of the leaf remaining green, which can progress and increase in size while remaining isolated, and in high density cause the tips to dry out and subsequently the leaf to burn downwards (WORDELL FILHO & BOFF, 2006).

Source: Grossklaus (2011)

**Figure 13**: Plants attacked by *Botrytis squamosa,* observed at EEItu, 2011.

Preventive measures can be taken by planting in places where there is no fog, avoiding lowland areas, avoiding mechanical damage, eliminating crop remains, treating seeds, rotating crops, using cover crops and resistant cultivars (SOUZA & RESENDE, 2002). Chemical control is the most widely used, especially in the seedbed phase and also in direct sowing. Various products are used for control and are presented in (Table 7).

### 2.10.3 Downy mildew - *Peronospora destructor* (Berk.). Casp.

It is considered the main disease of the onion crop, due to the speed with which the fungus spreads in the crop, which results in major losses (SOUZA & RESENDE, 2002). The pathogen is an obligate parasite and can survive on volunteer plants, bulbs and infected seeds. Wind is an efficient dispersal agent, as well as splashes of rain or irrigation water which can also contribute to dispersal (MAFFIA *et al.*, 2002).

The fungus develops well in mild temperatures and high humidity and is favored by prolonged periods of cloudy days. The optimum temperature for infection is between 10 and 20° C, and only a few hours (four hours) of leaf wetness are needed for the sporangia to germinate (MAFFIA *et al.,* 2002).

The symptoms appear as a large, oval, light green spot running lengthways, with violet-grey to dark mold (Figure 14), which is observed in the early hours of the morning (WORDELL FILHO & BOFF, 2006). As the lesions develop, the leaves can dry out and be attacked by saprophytic fungi such as Alternaria, as well as having their storage life compromised (MAFFIA *et al.*, 2002).

Source: Grossklaus (2011)

**Figure 14:** Onion plants attacked by *Peronospora destructor,* Ituporanga-SC Experimental Station, 2011.

Control of the disease should begin with eliminating the primary source of inoculum, using resistant cultivars that are more waxy, avoiding dense plantings, controlling irrigation and balanced fertilization. Most producers use chemical control through fungicides (Table 7) to control the disease.

### 2.10. 4Purple spot - *Alternaríaporri* (Ellis) Cif.

This disease is widespread in all onion-producing regions, causing severe damage and reducing production by up to 50-60% (SOUZA & RESENDE, 2002). The climatic conditions necessary for the disease to develop are high temperatures (18 to 30°) and relative humidity of over 90% (EPAGRI, 2000). In bulb crops, this disease is found mainly in late plantings (WORDELL FILHO & BOFF, 2006), characterized by favourable climatic conditions for the development of the disease.

The pathogen mainly attacks mature tissue and/or tissue infected by other pathogens (mainly mildew) and plants fertilized with excess nitrogen (WORDELL FILHO & BOFF, 2006) and attacked by *Thrips tabaci* (MAFFIA *et al.*, 2002). The initial symptoms are characterized by whitish to yellowish spots and turn wine-red (Figure 14) under conditions of high temperature and humidity (WORDELL FILHO & BOFF, 2006). Severely attacked leaves shrivel and wrinkle from the tip, developing into younger leaves and resulting in the production of smaller bulbs (SOUZA & RESENDE, 2002). The bulbs can also be infected at harvest and can cause rot on the outer scales (MAFFIA *et al.*, 2002).

Source: Grossklaus (2011)

**Figure 15:** Plants attacked by Purple Spot, after the mildew attack observed at the Ituporanga-SC Experimental Station, 2011.

Preventive measures should be taken to reduce the attack of the disease, such as crop rotation, irrigation control, avoiding injury and attack by mildew and thrips and proper nitrogen fertilizer management. The main products used for chemical treatment are shown in Table 7.

**Table 7**: Products used to control the main pests that attack the onion crop

| **Commercial Product** | **Active ingredient** | **Dosage Kg or lt/ha** | **Controlled diseases** |
|---|---|---|---|
| Sumilex 500WP | Promicidone | 1,0 a 1,5 | Botrytis cinerea; Purple Spot |
| Native* | Tebuconazole + Trifloxystrobin | 0,75 | Purple Spot |
| Folicur 200 EC* | Tebuconazole | 1,0 | Purple Spot |
| Sportak 480 EC* | Phrocloraz | 1,5 | Purple Spot |
| Rovral SC* | Iprodione | 1,5 | Purple Spot |
| Ridomil Gold MZ * | Metalaxyl-M + Mancozeb | 2,5 | Mildew |
| Unix 750 WG * | Ciprodinil | 0,375 | Purple Spot |
| Folio Gold | Metalaxyl-M + Chlorothalonil | 1,5 | Mildew |

Source: Agrofit (2011)

* Products used at EPAGRI - Ituporanga-SC Experimental Station

## 2.11 Harvesting, Storage and Marketing

The onion bulb is harvested when it reaches physiological maturity, which is manifested by the plant falling over or snapping due to the pseudostem wilting (EPAGRI, 2000). The percentage of popped plants varies, but it is recommended that the crop be harvested when it has 40 to 70% of yellowed or dried leaves, accompanied by toppling. When the bulbs are destined for long-term storage, they should be harvested when 50 to 80% of the plants are already cracked and the bulbs are ripe (BARBOSA, 2008).

Tipping over the remaining plants that have not popped naturally is practiced by a few producers in Santa Catarina (BARBOSA, 2008). Tumbling can be carried out when the snap does not occur naturally with the help of a wooden roller or water hose, thus avoiding the accumulation of water in the pseudostem of the plants, making the curing process faster and reducing post-harvest damage (EPAGRI, 2000).

Dry, sunny days are the best for harvesting; on the other hand, wet soil damages the preservation of the bulbs (SOUZA & RESENDE, 2002).

After the bulbs have been harvested, they must undergo a curing process, which can be natural or artificial. Curing is the phase in which the moisture content of the roots, husk and pescogo is markedly reduced immediately after harvesting (FINGER & CASALI, 2002).

Natural curing should be carried out in the field for 3 to 14 days, depending on the weather conditions, leaving the harvested plants on the ground, arranged in rows so that the leaves cover the bulbs to avoid sunburn or the formation of green pigmentation (EPAGRI, 2000).

Good curing is achieved when there is no rain or at most light rain after harvest (EPAGRI, 2000) and mild temperatures of around 24°C (BARBOSA, 2008).

Another practice used in the Alto Vale do Itajaí-SC region is pre-cutting. After harvesting, the plants remain in the field for 3-4 days, then the leaves are cut off, leaving approximately 5 cm of stalk, and the bulbs are packed in raffia bags, where they remain in the field until curing is complete, a process that takes an average of 10 days, after which they are stored or sold.

In addition to natural curing and pre-cutting, we also have the option of artificial curing. This practice is carried out by a minority of producers and consists of blowing heated or unheated air between the bulbs through perforated ducts at a temperature of 35 to 40°C for four to eight days. When done properly, it can speed up the curing process and produce quality bulbs, and the best quality is obtained by controlling humidity and temperature through the circulation of forced air (EPAGRI, 2000).

After curing, the bulbs can be sold or stored in warehouses arranged in racks, bins, bags and raffia sacks until they are ready for sale. Each cubic meter of onion bulbs with leaves has a mass of around 500 to 600 kg, a factor that helps those who want to size the storage area (EPAGRI, 2000).

When the producer deems it appropriate to market the bulbs, they should be processed by cutting off the pseudostem, leaving 1 cm, and trimming the roots without injuring the bulbs. They are then sorted by cross-sectional diameter (Table 8), eliminating bulbs that do not meet the standards set by MAPA and packed in bags with a capacity of 20 kg.

**Table 8:** Classification of onion bulbs according to the cross-sectional diameter defined by MAPA

| Class | Bulb cross-sectional diameter (mm) |
|---|---|
| 1 (lollipop) | < 35 |
| 2 | > 35 to 50 |
| 3 | >50 to 70 |
| 4 | >70 to 90 |
| 5 | > 90 |

According to the classification defined by the standard, the price received by the producer is equivalent to 100% for classes 3 and 4, 50% for class 2 and 30% for class 1, after the bulbs have been classified and processed.

# Chapter 3

## 3 FERTILIZATION EXPERIMENT: FERTILIZER DOSES PHOSPHATE FOR ONION CROPS AT DIFFERENT FERTILITY LEVELS

### 3.1 Introducing

The increase in the yield of the onion crop, as well as the health and quality of the bulbs, is greatly influenced by the availability of nutrients. According to MAGALHÁES (1993), the most absorbed elements in terms of dry matter percentage in onions are potassium (K) and nitrogen (N). In terms of nutrient concentration in the plant, at the different growth stages nitrogen, phosphorus and calcium reach a maximum at 130 days and decrease as the plant reaches the end of the cycle (SOUZA & RESENDE, 2002).

According to SOUZA & RESENDE (2002), phosphorus is the fourth nutrient in order of extraction and the one that offers the highest yields and increased bulb weight. According to MAGALHÁES (1993), the response to the dose of P applied depends entirely on the phosphorus content in the soil, the aluminum and iron content, as well as the pH and the amount of fertilizer applied. The P content in soils is between 0.2 and 5.0 g $kg^{-1}$ and only a small fraction is in available forms (ARAÚJO & MACHADO, 2006). According to MALAVOLTA *et al*, (2002) due to the process of fixation in the year of application the
P utilization is between 5% (in clay soils rich in Fe and Al oxides) and 25%, with the remainder being utilized in subsequent crops.

Responses to phosphate fertilization have been reported by different authors, which vary according to soil type, soil P content and quantities supplied. In order to find a response to the application of P to the onion crop in increasing doses, an experiment was carried out to evaluate the most economically efficient application of this element at different levels of natural fertility, following the recommendations for the region.

### 3.2 Materials and Methods

The work was carried out in the 2010/2011 and 2011/2012 harvests in the same experimental area (Experiment "A"), at EPAGRI - Ituporanga Experimental Station, located in the Alto Vale do Itajaí-SC region, specifically in the town of Águas Negras. The soil is classified as Cambissolo Húmico, with 30% clay, and the following chemical characteristics before the experiment was set up: pH-H2O=6.0; organic matter=3.2 %; P-Mehlich= 4.0 $mg/dm^3$ ; K= 62 $mg/dm^3$ ; Ca= 7.0 $cmolc/dm^3$ ; Mg= 3.2 $cmolc/dm^3$ ; CTC pH-7= 13.8 $cmolc/dm^3$ and base saturation (V%)= 74.9. In the 2011/2012 harvest, the same design was used (Experiment "B") in soil with different chemical characteristics: pH-H2O= 5.9; M.O=3.0% ; P-Mehlich= 26.5 $mg/dm^3$ ; K= 89$mg/dm^3$ ; Ca= 5.7 $cmolc/dm^3$ ; Mg=3.1 $cmolc/dm^3$ ; CTC pH-7= 13.9 $cmolc/dm^3$ and Base saturation (V%)= 64.4. Before the experiment was set up, turnip rape was grown as a green manure. When the turnip was in full bloom, it was managed with a knife roller, making it possible to transplant the onion seedlings by hand.

The experimental design was randomized blocks with six doses of P (0, 50, 100, 200, 400, 800 kg $ha^{-1}$ ) and four replications. The doses of phosphorus were applied before the seedlings were transplanted, distributed at random over the plots ($12m^2$ ) according to the treatment. The source of P used was triple super phosphate (Ca ($H2PO^-$ 4) 2H2O) with 42% P2O5. Potassium was applied at a dose of 90 kg $ha^{-1}$ of K2O, 60 kg at planting and 30 kg divided into two top dressings at 65 and 85 days after transplanting (DAT), using potassium chloride (KCl) as the K source. Ammonium nitrate ($NH4^+$ $NO3^-$ ) with 32% N was used as the nitrogen fertilizer, applying a total of 120 kg $ha^{-1}$ , of which 15 kg was distributed at planting and the rest as top dressing at 40, 65, 85 and 100 DAT, at doses of 30, 35, 20 and 20 kg $ha^{-1}$ respectively. Two applications of calcium nitrate (CaNO3) at 0.6%, boric acid (B3HO3) at 0.3% and zinc sulphate (ZnSO4) at 0.4% were made as foliar fertilization on 28/07 (13 DAT) and 05/08 (21 DAT).

The EMPASC 352-Bola Precoce cultivar was used, sown in April and transplanted on 15/07 in experiment "A" and 19/07 in experiment "B", with spacing of 0.36m between rows and 0.10m between plants, giving a total population of 277,000 plants/ha.

Weeds, pests and diseases were controlled using chemical products registered with MAPA. For weeds, two applications of herbicides were made: the first on 12/08 with a mixture of ioxynil (0.16 L $ha^{-1}$ ) and pendimenthalin (1.08 L $ha^{-1}$ ) and on 03/10 with a mixture of ioxynil (0.268 L $ha^{-1}$ ) and bentazon (0.18 L $ha^{-1}$ ) and weeding on 15/09. For pest control, insecticides were used on 28/07 chloripyrifos (0.288 L $ha^{-1}$ ), on 14/10 lambda-cyhalothrin (0.005 L $ha^{-1}$ ), on 21/10 chloripyrifos (0.72 L $ha^{-1}$ ) and on 31/10 lambda-cyhalothrin (0.005 L $ha^{-1}$ ). Seven applications were made to control fungal diseases (*Botrytis squamosa, Peronospora destructor and Alternaria porri*): a mixture of prochloraz ( 0.378 L $ha^{-1}$ ), metalaxyl (0.058 kg $ha^{-1}$ ) and mancozeb (0.92 kg $ha^{-1}$ ) was applied on 22/08; on 01/09 a mixture of metalaxyl (0.08 kg $ha^{-1}$ ), mancozeb (1.28 kg $ha^{-1}$ ), Tebuconazole ( 0.16 L $ha^{-1}$ ) and trifloxystrobin (0.08 L $ha^{-1}$ ); on 14/09 a mixture of metalaxyl (0.08 kg $ha^{-1}$ ), mancozeb (1.28 kg $ha^{-1}$ ) and Tebuconazole (0.2 L $ha^{-1}$ ); on 28/09 a mixture of metalaxyl (0.08 kg $ha^{-1}$ ), mancozeb (1.28 kg $ha^{-1}$ ) and Iprodione (0.75 L $ha^{-1}$ ); on 14/10 a mixture of metalaxyl (0.08 kg $ha^{-1}$ ), mancozeb (1.28 kg $ha^{-1}$ ) and Tebuconazole (0.2 L $ha^{-1}$ ); on October 21 a mixture of metalaxyl (0.08 kg $ha^{-1}$ ), mancozeb (1.28 kg $ha^{-1}$ ) and Tebuconazole (0.2 L $ha^{-1}$ ); and on October 31 a mixture of metalaxyl (0.08 kg $ha^{-1}$ ), mancozeb (1.28 kg $ha^{-1}$ ) and Iprodione (0.75 L ha ).$^{-1}$

On October 19, leaves were also collected from a total of 10 plots (from the center to the base, the last fully expanded leaf) for leaf analysis. Two sprinkler irrigations were also carried out (on October 6th and November 4th).

On 10/11, the plants that didn't pop naturally were toppled. Bulb harvesting and plant counting began on 18/11. The bulbs were pre-cured in the field. The bulbs were collected in plastic boxes, sorted and stored on 28/11, after which they were counted and weighed in order to obtain the yield. A post-harvest evaluation was carried out in April to determine post-harvest losses after the storage period according to the treatments.

### 3.3 Results and Discussion

The doses of phosphorus applied to the soil with P-Mehlich=4.0 mg/$dm^3$ significantly influenced the production of onion bulbs in the 2010/2011 harvest. The maximum yield of commercial bulbs was obtained with a dose of 800 kg $ha^{-1}$ of P2O5, which provided an increase of 20,098 kg $ha^{-1}$ (100.31%) compared to the control which received no phosphate fertilization (Figure 16A). There was also a significant difference in average bulb weight and plant height (data not shown). The doses of P tested did not influence the percentage of losses, the number of crooked and erect leaves, the average diameter of the bulbs and the number of leaves per plant (Table 9).

**Table 9:** Influence of the P2O5 dose on the biometric variables evaluated in the aerial part, as well as on the percentage of onion bulb losses in the 2010/2011 harvest.

| Treatment Doses of P2O5 | Total losses (%) | Number of erect leaves | Number Crooked leaves | Average diameter (cm) | Number of leaves per plant |
|---|---|---|---|---|---|
| **0** | 18,12 | 5,6 | 5,6 | 15,3 | 8,5 |
| **50** | 17,52 | 7,1 | 7,1 | 16,1 | 9,9 |
| **100** | 18,22 | 6,0 | 6,0 | 15,7 | 8,9 |
| **200** | 23,50 | 6,1 | 6,1 | 16,6 | 8,8 |
| **400** | 22,45 | 6,7 | 6,7 | 17,1 | 9,2 |
| **800** | 27,15 | 7,0 | 7,0 | 18,6 | 9,5 |
| F. | n.s. | n.s. | n.s. | n.s. | n.s. |

NOTE: n.s.: not significant. Untransformed data /Analysis: SASM-Agri

The plants that did not receive phosphate fertilization had a high percentage of bulbs belonging to class "2" (35 to 50mm in diameter), with an overall average of 18.65%. Bulbs in this class receive only 50% of the price paid for bulbs in class "3" (the highest price received). On the other hand, the number of class "2" bulbs in the treatments that received phosphate fertilization fell as the dose of added P increased (Figure 16B).

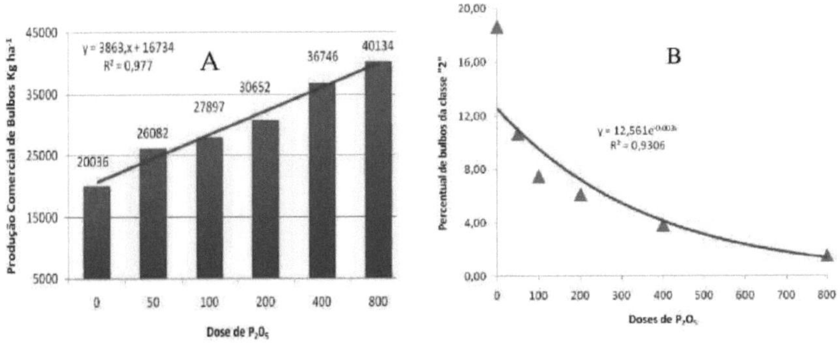


**Figure 16** Effect of P2O5 doses on the total yield of commercial bulbs (A) and the percentage of bulbs belonging to "class 2" (B) in an experiment conducted at EPAGRI-EEITU in the 2010/2011 harvest, in soil with P-Mehlich= 4.0 mg/dm$^3$ .

In the experiment carried out in the 2011/2012 harvest, the doses of P applied to the soil with P-Mehlich=4.0 mg/dm$^3$ also influenced commercial bulb production and the percentage of class "2" bulbs in a similar way to the 2010/2011 harvest. The maximum yield was obtained with the application of 800 kg ha$^{-1}$ of P205, a dose which provided an increase of 16,099 kg ha$^{-1}$ , i.e. 44.85% higher than the yield obtained for the control (Figure 17A), influencing yield in a polymonthly way according to the doses of P. The percentage of class "2" bulbs was reduced in a polymonthly way, with the maximum percentage for the plots which did not receive P fertilization (Figure B).

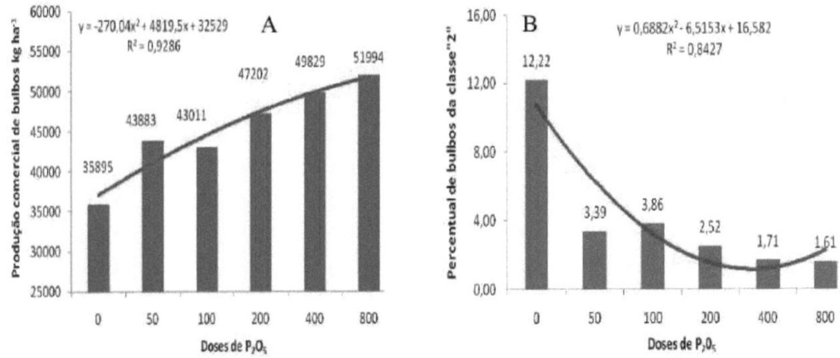


**Figure 17** Commercial yield of onion bulbs kg ha$^{-1}$ (A) and percentage of class "2" bulbs, according to the dose of P added to the soil at EPAGRI-EEITU in the 2011/2012 harvest, experiment "A", with P-Mehlich= 4.0 mg/dm$^3$ .

For experiment "B" in the 2011/2012 harvest with soil containing 26.5mg/dm$^3$ of P, there was no significant difference in the commercial production of bulbs (Figure 18), in the percentage of class "2" bulbs and in the other variables analyzed. This is mainly due to the balance between P-solid and P-solution. Most of the P added goes into P-solid from the soil, passing into P-solution as soon as the plants absorb it.

According to BISSANI *et al*, (2008) most of the P added to the soil goes into solid P, which is released into P-solution as the plants absorb P, and this is in equilibrium. Soils with levels considered high or very high according to the Manual of Liming and Fertilizing Recommendations for the States of RS and SC, the maximum dose to be applied would be the replacement value. On the other hand, in soils with low levels, P addition up to the maximum doses applied responded to P application, with the most economically efficient dose being around 200 kg $ha^{-1}$ .

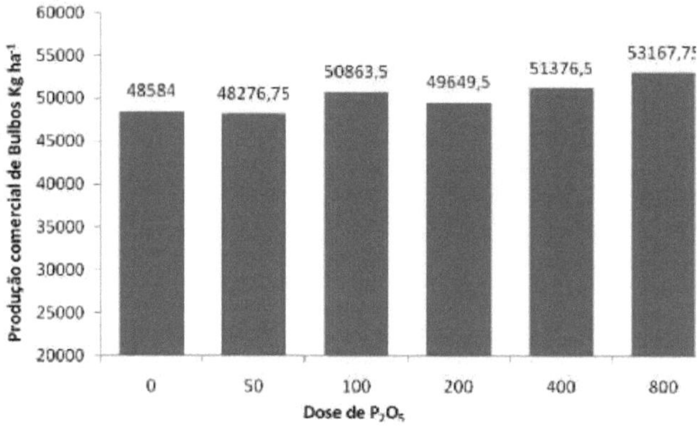


**Figure 18** Commercial production of onion bulbs kg $ha^{-1}$ according to the dose of P added to the soil at EPAGRI-EEITU in the 2011/2012 harvest, experiment "B", with P-Mehlich= 26.5mg/$dm^3$

# Chapter 4

## 4 FERTILIZER EXPERIMENT: FERTILIZER DOSES POTASSIUM FOR ONION CROPS

### 4.1 Introducing

The increase in yield of the onion crop, as well as the health and quality of the bulbs, is greatly influenced by the availability of nutrients. The most absorbed elements in terms of dry matter percentage in onions are potassium (K) and nitrogen (N) (MAGALHAES, 1993). K requirements for optimum plant growth are in the range of 20-50 g $kg^{-1}$ of the mass of the dry vegetative parts of the plant, fruit and tubers (MEURER, 2006).

Although onions extract a large amount of potassium, responses to this nutrient have generally not been observed, although it can improve bulb quality, resistance to certain leaf diseases and post-harvest preservation (SOUZA & RESENDE, 2002). An increase in productivity was seen by (RESENDE *et al.* 2008) with an application of 90 kg $ha^{-1}$ of $K_2O$, and a lower quantity of bulbs in Class "0" was achieved up to a dose of 150 kg $ha^{-1}$ of $K_2O$ (MAY *et al.* 2007). The appropriate amount of potassium for best preservation and quality assurance was 90 kg $ha^{-1}$ of $K_2O$, as observed by RESENDE *et al.* (2008).

Increased productivity is not always achieved, especially when the soil already has good potassium availability, as well as its effect on the quality and preservation of the bulbs. With the aim of evaluating the response of onions to potassium application, an experiment was set up to determine the most economically efficient dose, as well as the quality and post-harvest preservation of onion bulbs.

### 4.2 Materials and Methods

The work was carried out during the 2010/2011 and 2011/2012 harvests at EPAGRI - Ituporanga Experimental Station, located in the Alto Vale do Itajaí-SC region. The soil is classified as Cambissolo Húmico, with 30% clay and had the following chemical characteristics before the experiment was set up: $pH_{H2O}$=6.0; M.O=3.2%; P- Mehlich=4.0mg/$dm^3$ ; K=62mg/$dm^3$ ; Ca=7.0cmolc/$dm^3$ ; Mg=3.2cmolc/$dm^3$ ; CTC pH- 7=13.8cmolc/$dm^3$ and Base Saturation (V%)=74.9. In the 2010/11 harvest, the conventional soil management system was used and in the 2011/12 harvest, the minimum cultivation system was used on spontaneous summer plants and in both harvests the seedlings were transplanted manually.

The experimental design was randomized blocks, with six doses of K (0, 50, 100, 200, 400, 800 kg $ha^{-1}$ ) and four replications, on the same experimental area as the 2010/2011 harvest. Potassium (K) was applied according to the treatment, with potassium chloride (KCl) applied by hand as the K source. Phosphate fertilizer was applied before transplanting, and the fertilizer was distributed by hand on the plots (12$m^2$ ), using triple super phosphate (Ca $(H2PO^-{}_4)_2 2H_2O$) with 42% P2O5 at a dose of 200kg $ha^{-1}$ . Ammonium nitrate ($NH_4^+$ $NO_3^-$ ) at 32% was used as a nitrogen fertilizer, with a total dose of 120 kg $ha^{-1}$ , 15 kg of which was applied at planting and the rest in top dressing at 40, 65, 85 and 100 DAT, at doses of 30, 35, 20 and 20 kg $ha^{-1}$ respectively. As a foliar fertilizer, two applications were made of calcium nitrate (Ca $NO_3$) at 0.6%, boric acid (B3 $HO_3$) at 0.3% and zinc sulphate (Zn $SO_4$) at 0.4% in a mixture on 28/07 (13 DAT) and 05/08 (21 DAT).

The EMPASC 352-Bola Precoce cultivar was used, sown in April and transplanted with seedlings at the thickness of a pencil, with a spacing of 0.36m between rows and 0.10m between plants, thus obtaining 277,000 plants/ha. The seedlings were transplanted manually.

Weeds, pests and diseases were controlled using chemical products registered with MAPA. For weeds, two applications of herbicides were made: the first on 12/08 with a mixture of ioxynil (0.16 L $ha^{-1}$ ) and pendimenthalin (1.08 L $ha^{-1}$ ) and on 03/10 with a mixture of ioxynil (0.268 L $ha^{-1}$

) and bentazon (0.18 L $ha^{-1}$ ) and weeding on 15/09. For pest control, insecticides were used on 28/07 chloripyrifos (0.288 L $ha^{-1}$ ), on 14/10 lambda-cyhalothrin (0.005 L $ha^{-1}$ ), on 21/10 chloripyrifos (0.72 L ha-1) and on 31/10 lambda-cyhalothrin (0.005 L $ha^{-1}$ ). Seven applications were made to control fungal diseases (*Botrytis squamosa, Peronospora destructor and Alternaria porri*): the first on 22/08 with a mixture of prochloraz ( 0.378 L $ha^{-1}$ ), metalaxyl (0.058 kg $ha^{-1}$ ) and mancozeb (0.92 kg $ha^{-1}$ ); on 01/09 a mixture of metalaxyl (0.08 kg $ha^{-1}$ ), mancozeb (1.28 kg $ha^{-1}$ ), Tebuconazole (0.16 L $ha^{-1}$ ) and trifloxystrobin (0.08 L $ha^{-1}$ ); on 14/09 a mixture of metalaxyl (0.08 kg $ha^{-1}$ ), mancozeb (1.28 kg $ha^{-1}$ ) and Tebuconazole (0.2 L $ha^{-1}$ ); 28/09 mixture of metalaxyl (0.08 kg $ha^{-1}$ ), mancozeb (1.28 kg $ha^{-1}$ ) and Iprodione (0.75 L $ha^{-1}$ ); 14/10 mixture of metalaxyl (0.08 kg $ha^{-1}$ ), mancozeb (1.28 kg $ha^{-1}$ ) and Tebuconazole (0.2 L $ha^{-1}$ ); 21/10 mixture of metalaxyl (0.08 kg $ha^{-1}$ ), mancozeb (1.28 kg $ha^{-1}$ ) and Tebuconazole (0.2 L $ha^{-1}$ ); and 31/10 mixture of metalaxyl (0.08 kg $ha^{-1}$ ), mancozeb (1.28kg $ha^{-1}$ ) and Iprodione ( 0.75 L ha ).$^{-1}$

On October 19, a total of 10 leaves were collected (from the center to the base, the last fully expanded leaf) for leaf analysis of nutrient content. There were only two sprinkler irrigations throughout the cycle (on October 6th and November 4th).

On November 10, the plants that didn't pop naturally were toppled and the bulbs were harvested on November 18. After the plants had been pruned, they were left in the field to pre-cure. After pre-curing, the bulbs were counted and weighed in order to obtain the yield. The bulbs were collected in plastic boxes, sorted and stored on 28/11. A post-harvest evaluation will be carried out in April to determine the yield after storage according to the treatments.

### 4.3 Results and Discussion

As for the response of onions to potassium application in the 2010/2011 harvest, it can be seen that there was no significant difference between the treatments, either in yield (Figure 19A) or in post-harvest losses (Figure 19B). Several authors have reported that plants well-nourished in K are better preserved in post-harvest, storage and transportation, and no interaction was observed for these variables. Plant height and average leaf diameter also did not interact between treatments (Table 10).

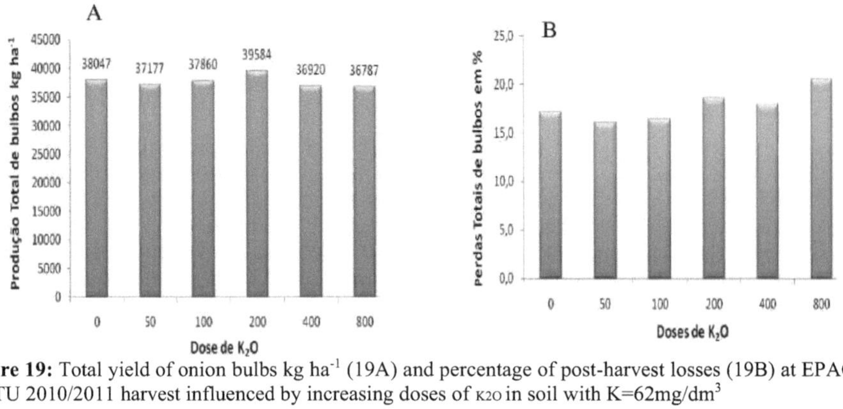


**Figure 19:** Total yield of onion bulbs kg $ha^{-1}$ (19A) and percentage of post-harvest losses (19B) at EPAGRI-EEITU 2010/2011 harvest influenced by increasing doses of $K_2O$ in soil with K=62mg/$dm^3$

**Table 10:** Influence of potassium fertilization on the onion crop at EPAGRI-EEITU for the 2010/2011 harvest.

| Dose of K2O | Plant height (cm) | Diameter Average leaf size (mm) | Class "2" bulbs (%) |
|---|---|---|---|
| 0 | 71,88 | 18,35 | 1,82 |
| 50 | 71,73 | 17,70 | 3,02 |
| 100 | 72,20 | 18,13 | 3,05 |
| 200 | 74,05 | 17,65 | 1,80 |
| 400 | 70,95 | 17,35 | 2,14 |
| 800 | 73,48 | 18,43 | 3,17 |
| | n.s | n.s | n.s |

OBS.: data analyzed by SASM-Agri; n.s.: not significant

Similar to the previous year's crop, in the 2011/2012 crop commercial bulb production was not influenced by the doses of K applied (Figure 20). In this season, the application of this element also did not influence the other variables analyzed, as can be seen in (Table 11). However, it was observed in the field that the plots that did not receive potassium fertilization showed yellowing of the older leaves and drying of the tips, in agreement with HAAG *et al.* (1968), who found no reduction in the average size of the bulbs. Post-harvest losses from the 2011/2012 harvest will only be assessed in April 2012, so it will not be possible to analyze this data.

One of the possible causes of the lack of response to the application of this element may be related to the release of K from primary minerals found in the region, since the amount absorbed by the onion crop is often greater than the amounts found in the soil analysis results.

**Table 11:** Influence of potassium fertilization on the onion crop at EPAGRI-EEITU 2011/2012 harvest.

| Dose of K2O | Rotten bulbs (%) | Average bulb weight (mm) | Class "2" bulbs (%) |
|---|---|---|---|
| 0 | 0,38 | 178 | 2,15 |
| 50 | 0,96 | 177 | 3,01 |
| 100 | 1,58 | 184 | 2,17 |
| 200 | 1,64 | 192 | 1,77 |
| 400 | 2,3 | 184 | 2,38 |
| 800 | 0,98 | 189 | 2,27 |
| **Results** | **n.s.** | **n.s.** | **n.s.** |

NOTE: Data analyzed by SASM-Agri; ns: not significant

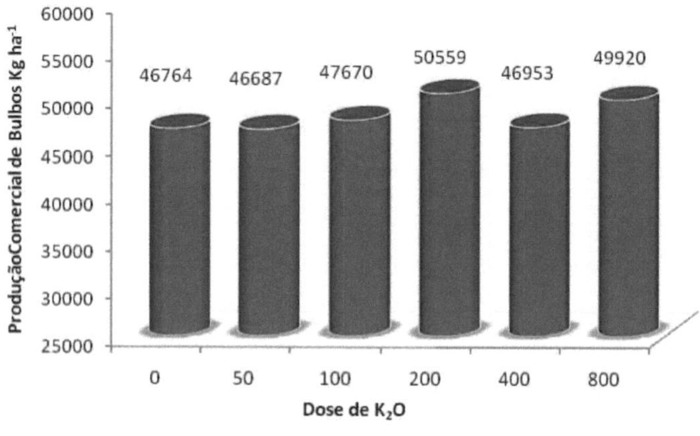


**Figure 20:** Commercial production of onion bulbs (kg $ha^{-1}$ ) at EPAGRI-EEITU 2011/2012 harvest influenced by increasing doses of $K_2O$ in soil with K=62mg/$dm^3$

According to the results obtained for the two consecutive harvests, it was observed that the application of K, in soils with medium and high levels of fertility, does not influence the commercial production of bulbs, nor the percentage belonging to class "2", and it is recommended to apply the amount of fertilizer referring to the replacement value recommended in the Manual of Liming and Fertilizing Recommendations for the States of RS and SC without the need for high doses.

# Chapter 5

## 5 FERTILIZER EXPERIMENT: FERTILIZER DOSES ORGANOMINERALS FOR ONION CROPS

### 5.1 Introducing

The increase in yield of the onion crop, as well as the health and quality of the bulbs, is greatly influenced by the availability of nutrients, which can be obtained from mineral fertilizers or organic fertilizers. The onion crop responds very well to organic fertilizer, which, as well as improving fertility, acts beneficially on the physical and biological conditions of the soil (EPAGRI, 2000). As well as supplying nutrients, organic fertilizers increase the organic matter content, improve the structure and aeration of the soil and its ability to store moisture. They also act as a soil temperature regulator, slow down phosphorus fixation and increase the Cation Exchange Capacity (CEC), preventing nutrient losses through leaching (MALAVOLTA *et al,* 2002).

If sufficient organic material is available, it is possible to supply the nutrients needed for the growth and production of most crops, including onions. However, it is necessary to evaluate the doses, the material used, its interaction with climatic conditions and its use as a complement to mineral fertilization. Looking for an answer to organic fertilization in comparison with mineral fertilization, an experiment was carried out in the 2010/2011 and 2011/2012 harvests to evaluate the production and post-harvest conservation of onion crops, since the manual recommending liming and fertilization for the states of Santa Catarina and Rio Grande do not contain specific organic fertilization recommendations for these crops.

### 5.2 Materials and Methods

The work was carried out during the 2010/2011 and 2011/2012 harvests at EPAGRI's Ituporanga Experimental Station, located in the Alto Vale do Itajaí region of Santa Catarina. The soil is classified as Cambissolo Húmico, with 33% clay, and had the following chemical characteristics before the experiment was set up: pH-H2O=6.2; M.O=3.0%; P- Mehlich=32.0 mg/dm$^3$ ; K=232 mg/dm$^3$ ; Ca=7.0 cmolc/dm$^3$ ; Mg=3.1 cmolc/dm$^3$ ; CTC pH- 7=13.4 cmolc/dm$^3$ and Bases saturation (V%)= 79.5. In the 2010/2011 harvest, the conventional soil management system was used and in the 2011/2012 harvest, the minimum tillage system was used on spontaneous summer plants. In both harvests, the seedlings were transplanted manually.

The experimental design adopted was a randomized block design, in which six treatments were tested, consisting of different amounts and sources of nutrients, with four replications. Treatments T1, T2, T3 and T4 used (0.0; 4.5; 9.0; and 18.0 Mg ha$^{-1}$ of poultry manure) respectively, distributed by spreading before transplanting the seedlings. In T5, mineral fertilizer was applied as recommended in the Fertilizing and Liming Manual (CQFS - RS/SC, 2004), 100, 120 and 90 kg ha$^{-1}$ , for N, P and K, respectively. In treatment T6, 4.5 Mg ha$^{-1}$ of poultry manure + mineral supplementation with N and P (46 and 62.4 kg ha$^{-1}$ ) were applied according to the levels shown by the chemical analysis of the soil and the Fertilization and Liming Manual (CQFS - RS/SC, 2004).

In treatments T5 and T6, KCl (potassium chloride) was used as the source of potassium, which was applied by spreading it on the surface of the experimental plots. A similar procedure was also adopted for phosphate fertilization, but triple super phosphate (Ca (H2PO$^-$ 4) 2H2O) with 42% P2O5 was used as the P source. Ammonium nitrate ($NH_4^+$ $NO_3^-$ ) was used as the nitrogen fertilizer in treatment T5, with 32%, of which 120 kg ha$^{-1}$ was applied 15 kg at planting and the rest in top dressing at 40, 65, 85 and 100 DAT, at doses of 30, 35, 20 and 20 kg ha$^{-1}$ respectively. The foliar fertilizer was applied twice with calcium nitrate (Ca $NO_3^-$ ) at 0.6%, boric acid (B3 HO3) at 0.3% and zinc sulphate (Zn SO4) at 0.4% in a mixture applied 13 and 21 days after transplanting.

The EMPASC 352-Bola Precoce cultivar was used, sown in April and transplanted with seedlings at the thickness of a pencil, with spacing of 0.36m between rows and 0.10m between plants, thus obtaining 277,000 plants/ha. The seedlings were transplanted manually.

Weeds, pests and diseases were controlled using chemical products registered with MAPA. For weeds, two applications of herbicides were made: the first on 12/08 with a mixture of ioxynil (0.16 L $ha^{-1}$ ) and pendimenthalin (1.08 L $ha^{-1}$ ) and on 03/10 with a mixture of ioxynil (0.268 L $ha^{-1}$ ) and bentazon (0.18 L $ha^{-1}$ ) and weeding on 15/09. For pest control, insecticides were used on 28/07 chloripyrifos (0.288 L $ha^{-1}$ ), on 14/10 lambda-cyhalothrin (0.005 L $ha^{-1}$ ), on 21/10 chloripyrifos (0.72 L $ha^{-1}$ ) and on 31/10 lambda-cyhalothrin (0.005 L $ha^{-1}$ ). Seven applications were made to control fungal diseases (*Botrytis squamosa, Peronospora destructor and Alternaria porri*): the first on 22/08 with a mixture of prochloraz ( 0.378 L $ha^{-1}$ ), metalaxyl (0.058 kg $ha^{-1}$ ) and mancozeb (0.92 kg $ha^{-1}$ ); on 01/09 a mixture of metalaxyl (0.08 kg $ha^{-1}$ ), mancozeb (1.28 kg $ha^{-1}$ ), Tebuconazole (0.16 L $ha^{-1}$ ) and trifloxystrobin (0.08 L $ha^{-1}$ ); on 14/09 a mixture of metalaxyl (0.08 kg $ha^{-1}$ ), mancozeb (1.28 kg $ha^{-1}$ ) and Tebuconazole (0.2 L $ha^{-1}$ ); 28/09 mixture of metalaxyl (0.08 kg $ha^{-1}$ ), mancozeb (1.28 kg $ha^{-1}$ ) and Iprodione (0.75 L $ha^{-1}$ ); 14/10 mixture of metalaxyl (0.08 kg $ha^{-1}$ ), mancozeb (1.28 kg $ha^{-1}$ ) and Tebuconazole (0.2 L $ha^{-1}$ ); 21/10 mixture of metalaxyl (0.08 kg $ha^{-1}$ ), mancozeb (1.28 kg $ha^{-1}$ ) and Tebuconazole (0.2 L $ha^{-1}$ ); and 31/10 mixture of metalaxyl (0.08 kg $ha^{-1}$ ), mancozeb (1.28kg $ha^{-1}$ ) and Iprodione ( 0.75 L ha ).$^{-1}$

On October 19, a total of 10 leaves were collected (from the center to the base, the last fully expanded leaf) for leaf analysis of nutrient content. There were only two sprinkler irrigations throughout the cycle (on October 6th and November 4th).

Harvesting of the bulbs began on November 24th when 70% of the plants were popped and left in the field to pre-cure. After pre-curing, the bulbs were counted and weighed in order to obtain the yield. The bulbs were collected and stored on December 5 in plastic boxes where they will remain until April to assess post-harvest losses.

### 5.3 Results and Discussion

Figure 21 shows that for the 2010/2011 harvest, the addition of nutrients influenced total bulb production, for which when considering all the treatments, a polynomial behavior was obtained. The yield obtained in the control plots (which received no fertilizer) was 30.89% lower than the yield obtained in the plots where mineral fertilizer was applied (T5) and 57.62% lower than the yield obtained in the plots where the highest dose of poultry litter was applied (T4). The different forms and doses of fertilizer also influenced the percentage of class "2" bulbs, with a total of 10.48% of bulbs belonging to this class being obtained for the control, which has a 50% lower nail than class "3" bulbs. The number of leaves and plant height variables were also statistically lower than those obtained for the other treatments (Table 12).

**Table 12:** Influence of different sources and doses of nutrients on the development of onions grown at EPAGRI-EEITU during the 2010/2011 harvest. Treatments T1 - no fertilization; T2, T3 and T4 - application of 4.5, 9.0 and 18.0 Mg $ha^{-1}$ of poultry manure; T5 - mineral fertilization (NPK) and T6 - 4.5 Mg of poultry manure + supplementary mineral fertilization.

| Treatment | Average bulb weight (g) | Plant height (cm) | Storage losses (%) | Number of leaves per plant |
|---|---|---|---|---|
| T1 | 116,25 d | 63,53 c | 14,05 b | 8,06 c |
| T2 | 153,25 c | 71,13 b | 19,55 b | 9.43 ab |
| T3 | 176.25 ab | 73.23 ab | 22.48 ab | 9.44 ab |
| T4 | 184,50 a | 77,30 a | 32,40 a | 9.78 ab |
| T5 | 160.00 bc | 74.15 ab | 18,95 b | 9,03 b |
| T6 | 164.75 bc | 76.28 ab | 22.00 ab | 9.83 ab |

Means followed by the same letter in the column do not differ according to the Tukey test (SASM-Agri)

When using poultry litter, the highest yield of commercial bulbs was obtained with the application of 18 Mg $ha^{-1}$ , but this was 20.5% higher than the yield obtained with the lowest dose of litter tested and 17.08% higher than the yield obtained with the application of mineral fertilizer (T5), as can be seen in Figure 21. For treatment T6 (organic fertilizer plus mineral supplementation), similar results were obtained to T5 (mineral fertilizer) and no statistical difference was observed between these treatments in the variables evaluated.

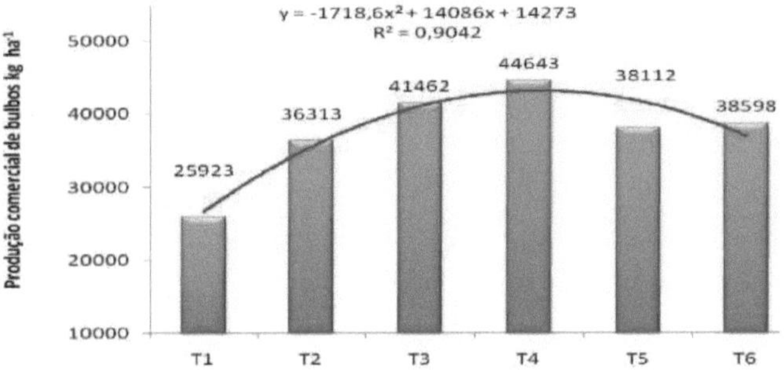


**Figure 21:** Total yield of onion bulbs (kg $ha^{-1}$ ) as a function of organic and organic-mineral fertilization at EPAGRI-EEITU 2010/2011 harvest for treatments T1- no fertilization; T2, T3 and T4 application of 4.5; 9.0; and 18.0 Mg $ha^{-1}$ of Poultry Manure); T5- Mineral Fertilization (NPK) and T6- 4.5Mg of Poultry Manure + complementary Mineral Fertilization.

For the variable total loss of bulbs after the storage period, the plots that received the highest dose of poultry litter (T4) showed the highest percentage of losses (32.4%), as can be seen in Figure 22. According to the results obtained, it is possible to recommend not using high doses of poultry litter, especially when the onion producer intends to store the bulbs for a certain period of time, which could lead to significant losses, thus reducing the economic viability of the crop. For producers who intend to store the bulbs after the curing process, the use of poultry litter with doses close to 9.0 Mg ha-1, or organomineral fertilizer, as well as mineral fertilizer can be used, since these treatments were where the lowest percentage of damaged bulbs was observed.

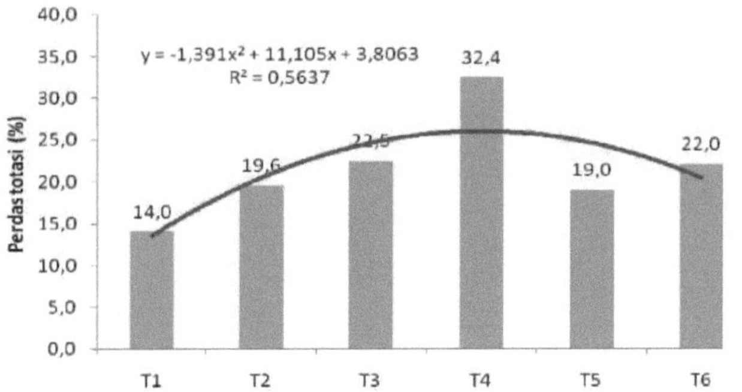


**Figure 22:** Percentage of post-harvest losses after the bulb storage period in the 2010/2011 harvest for treatments T1 - no fertilization; T2, T3 and T4 - application of 4.5; 9.0; and 18.0 Mg $ha^{-1}$ of Poultry Manure); T5 - Mineral Fertilization (NPK) and T6 - 4.5Mg of Poultry Manure + complementary Mineral Fertilization.

Organic fertilization up to a maximum dose of 9.0 Mg ha-1 of poultry manure or in addition to mineral fertilization is interesting for onion crops, mainly due to its characteristics, which according to MALAVOLTA *et al*, (2002), contributes to increasing the CEC and the availability of nutrients, as well as improving the chemical, physical and biological attributes of the soil. In addition, the increase in organic matter in the soil results in a reduction in the density of the surface layer and a consequent increase in aeration, drainage and water retention (BISSANI *et al.*, 2008).

In the 2011/2012 harvest, maximum production was obtained with mineral fertilization (T5), with totals of 18,924 kg ha$^{-1}$ (68.46%) higher than the control. Organo-mineral fertilization (T6) showed the same behavior, with 17,915 kg ha$^{-1}$ (64.81%) more than the control. Organic fertilization with 18 Mg ha$^{-1}$ (T4) compared to mineral fertilization (T5) and organomineral fertilization (T6) did not differ in terms of commercial production and percentage of class "2" bulbs (Figure 23 and 24). The highest percentages of class "2" bulbs were observed for the treatment that received no fertilizer.

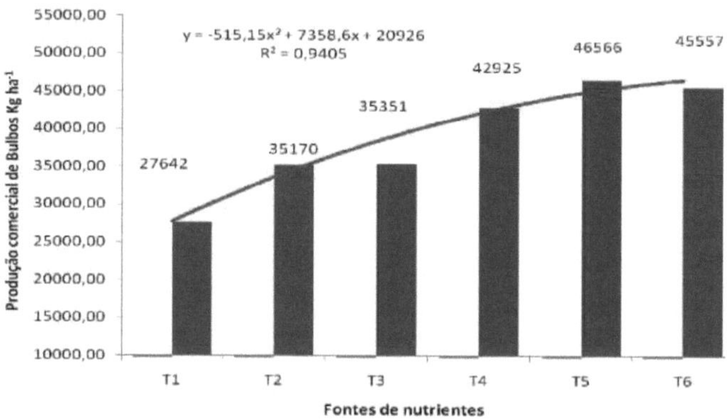


**Figure 23:** Commercial production of onion bulbs (kg ha-1) as a function of organic and organic-mineral fertilization at EPAGRI-EEITU during the 2010/2011 harvest for treatments T1- no fertilization; T2, T3 and T4 - application of 4.5, 9.0 and 18.0 Mg ha$^{-1}$ of poultry manure; T5 - mineral fertilizer (NPK) and T6 - 4.5 Mg of poultry manure + supplementary mineral fertilizer.

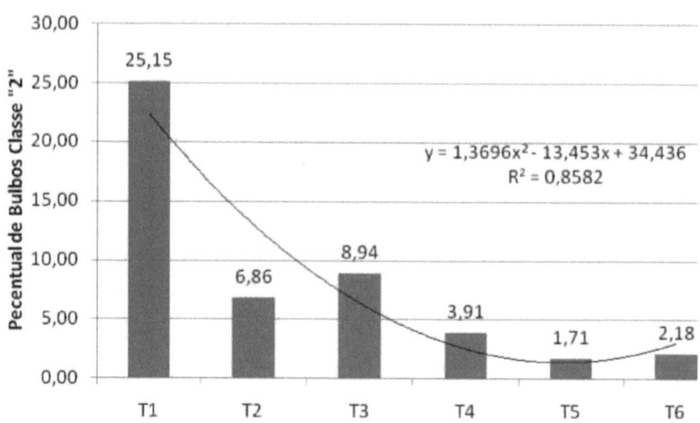


**Figure 24:** Percentage of class "2" bulbs in the 2011/2012 harvest as a function of organic and organic-mineral fertilization at EPAGRI-EEITU 2010/2011 harvest for treatments T1- no fertilization; T2, T3 and T4 application of 4.5; 9.0; and 18.0 Mg ha$^{-1}$ of Poultry Manure); T5- Mineral Fertilization (NPK) and T6- 4.5Mg of Poultry Manure + complementary Mineral Fertilization.

In the field, it was observed that the plots that received only mineral and organomineral fertilization showed earlier ripening when compared to the treatments with organic fertilization. This is possibly due to the climatic conditions that occurred this year, since the excess rainfall and leaching of nutrients in the plots with mineral fertilizer may have contributed to the shortening of the cycle in these treatments. On the other hand, in the plots where only organic fertilizer was applied, because the release of nutrients was slower and there was less leaching, the plants were better able to absorb them, thus extending the crop cycle.

In general, looking at the results obtained in the 2010/2011 and 2011/2012 harvests, there were differences in the commercial production of bulbs and class "2" bulbs. In years when rainfall is above the climatological average, organic fertilizer up to a dose of 18 Mg $ha^{-1}$ is not enough to meet the crop's needs, while in years when rainfall is normal, high yields of commercial bulbs are achieved, although these have a high rate of post-harvest losses. Organic fertilizer, complemented with mineral fertilizer, is an interesting alternative for fertilizing onion crops. Organo-mineral fertilization can be used not only to increase total bulb production, but also to improve the soil's physical and structural conditions over time by increasing organic matter and the soil's CEC and gradually releasing nutrients throughout the crop cycle.

# FINAL CONSIDERATIONS

Research has shown that the increase in the production of onion bulbs, as well as their quality and preservation, is entirely related to the availability of nutrients.

For phosphate fertilization, it was observed that in soils with low fertility and low P content, the onion responds to fertilization, with a significant increase in commercial production and in the percentage of class "2" bulbs, mainly due to the behaviour of P in the soil, which shows low efficiency in the year of application. In soils with a high P content, the response to fertilization was not verified, and it is therefore recommended to apply according to the Liming and Fertilization Manual for the states of Santa Catarina and Rio Grande do Sul, using the replacement value.

For potassium, there were no significant responses in commercial production, class "2" bulbs, post-harvest preservation and the other variables analyzed in the 2010/2011 and 2011/2012 harvests. One of the reasons for this behavior could be the release of K from the primary minerals by the weathering process, since the amount extracted by the onion crop is many times greater than the amount in the soil. Another possible cause for the lack of response to this element may be related to the extractor used to determine the level of potassium in the soil, since it may not be able to extract all the K in the soil.

In relation to organic fertilization, there were differences between the agricultural years evaluated, which may be mainly related to climatic conditions. In years with normal rainfall, organic fertilizer can be an alternative for onion nutrition, aiming for high bulb production. However, the bulbs will have to be marketed soon after harvest, since if they are stored, there will be a high percentage of losses. In years with above-average rainfall, there is no significant difference in bulb production when using organic, mineral and organo-mineral fertilizers. However, it should be pointed out that the use of organo-mineral fertilizer is an interesting alternative when good quality, low-cost organic material is available, in order to supply nutrients gradually and increase the organic matter content in the soil, thereby improving the cation exchange capacity, water retention and aeration of the soil, and consequently the availability of nutrients for the plants.

**Thanks**

Agradego,

To God;

My wife Marciane for her patience, trust and encouragement;

Family members for their support in every aspect;

My classmates, for their conviviality, friendship, fun and exchange of experiences;

Teachers for their encouragement and dedication;

To Professor Romano for his initiative, encouragement and persistence;

To the internship supervisor Claudinei for the opportunity and the experience he passed on;

To the course coordinator, Professor Marcuzzo, for his determination and trust;

To EPAGRI for granting the internship;

To the others who accompanied me on this journey and contributed in some way;

# BIBLIOGRAPHY


ARAUJO, A.P.; MACHADO, C.T.de T. Fósforo. SBCS, Vigosa, 2006. **MINERAL NUTRITION DE PLANTAS,** 432p. (ed. FERNANDES, M.S.)

BARBOSA, C. A. **MANUAL DA CULTURA DA CEBÓLA**. 1.ª Ed. - Vinosa, 2008. 149p. ISBN: 978-85-61250-01-0.

BISSANI, C. A; GIANELLO, C; CAMARGO, F. A. de O; TEDESCO, M. J. **SOIL FERTILITY AND CULTURE MANAGEMENT**. 2nd Ed. Porto Alegre: Metrópole, 2008.

BOEING, G. **FACTORS AFFECTING THE QUALITY OF BEET IN FAMILY AGRICULTURE IN CATARINA**. Florianópolis: Instituto Cepa/SC, 2002. 80p. ISBN 85-88974-03-7

CANTERI, M. G., ALTHAUS, R. A., VIRGENS FILHO, J. S., GIGLIOTI, E. A., GODOY, C. V. **SASM - AGRi : SYSTEM FOR ANALYSIS AND SEPARATION OF AVERAGES IN AGRICULTURAL EXPERIMENTS BY THE ScOFT - KNOTT, TuKEY AND DuNcAN METHODS**. Revista Brasileira de Agrocomputagão, V.1, N.2, p.18-24. 2001.

COSTA, Nivaldo Duarte *et al.* Onion cultivars. **iNFORME AGROPEcuÁRiO**. Belo Horizonte, v. 23, n. 218, p. 20-77, 2002.

Soil Chemistry and Fertility Commission - RS/SC. **MANUAL DE ADuBAcaO E DE cALAGEM PARA OS ESTADOS DO RiO GRANDE DO SuL E DE SANTA cATARiNA**. 10. ed. - Porto Alegre: Sociedade Brasileira de Ciencia do Solo, 2004. 400p.

EMATER-RS. BEET **CULTURE IN RiO GRANDE DO SuL**. Org. by Roque Ten Caten and Thomaz Lucia. Porto Alegre, 1982.

EMBRAPA. **CEBOLA CULTURE IN THE NORTHEAST**. Available at: http://sistemasdeproducao.cnptia.embrapa.br/FontesHTML/Cebola/CultivoCebolaNordeste/botanica.htm. Accessed on: August 17, 2011.

EPAGRI. **SÍNTESE DA AGRicuLTuRA cATARiNENSE 2009-2010**. Available at: http://cepa.epagri.sc.gov.br/Publicacoes/Sintese_2010/sintese%202010_inteira.pdf. Accessed August 16, 2011.

EPAGRI. **PRODUCTION SYSTEM FOR BEET**. Santa Catarina (3rd revision). Florianópolis: 2000 91p. (Epagri - Production Systems, 16).

EPAGRI/CIRAM. **CENTRO DE INFORMACOES DE RECURSOS AMBIENTÁIS E DE HiDROMETEOROLOGiA DE SANTA cATARiNA**. Available at: http://ciram.epagri.sc.gov.br/portal/website/index.jsp?url=jsp/agricultura/zoneAgroecologico.jsp&tipo=agricultura. Accessed November 05, 2011.

FAO. **FooD AGRicuLTuRAL oRganiZATioN**. Available at: http://faostat.fao.org/DesktopDefault.aspx?PageID=339&lang=es. Accessed December 15, 2011.

GONQALVES. P. A. de S. Ecological Management of the Main Onion Pests. In: WORDELL FILHO, J.A.; ROWE, E.; GONCALVES, P.A.S.; DEBARBA, J.D. **MANEJO FiTossANiTARIO DA cuLTuRA DA cEBoLA**. Florianópolis: Epagri, 2006. 226p. p. 168-189.

IPNI/BRASIL. **HOW TO MAKE A SOIL COLLECT**. Available at: http://www2.ipni.net/ppiweb/DiagNutricional.nsf/ColetaSolo?OpenPage. Accessed August 3, 2011.

KASSAB, Á. L. **ONION: FROM THE TOMB OF THE PHARAOHS TO DEMANDING MODERN TABLES**. Rio de Janeiro. Tecnoprint, 1987. 114p.

KURTZ, C. **ONION YIELD INFLUENCED BY THE ADDITION OF MICRONUTRIENTS AND NITROGEN**. 2008. 59p. Dissertation (Master's degree in soil management) Santa Catarina State University, Lages, 2008.

MAFFIA, L.A.; MIZUBUTI, E.S.G.; PEDROSA, R.A. Onion diseases. **INFORME AGRoPEcUÁRio**. Belo Horizonte, v. 23, n. 218, p. 75-87, 2002.

MAGALHAES, J.R. **DIAGNOSIS OF NUTRITIONAL DISORDERS IN VEGETABLES**. Brasília-EMBRAPA-DPU, 1988. 64p.

MAGALHAES, J.R. Onion Nutrition and Fertilization. Vegetable nutrition and fertilization. Anais / edited by M.E. Ferreira; P.D. Castellane; M.C.P. Cruz. Piracicaba**: PoTAFos**, 1993.

MALAVOLTA, E; VITTI, G.C.; OLIVEIRA, S.A.; Evaluation of plant nutritional status: Principles and Applications. Piracicaba: **ASSOCIAZÃO BRASILEIRA PARA PESQUISA DA**

**PoTAssA E Do FosFATo.** 1989. 201p.
MAP. AGROFIT- **AGRoTÓXIC FiTossANiTARY sysTEM.** Available at: http://extranet.agricultura.gov.br/agrofit_cons/principal_agrofit_cons. Accessed on: August 18, 2011.
MAP. NATIONAL **of cuLTiVAREs.** Available at: http://www.agricultura.gov.br/vegetal/registros-autorizacoes/registro/registro-nacional-cultivars. Accessed on: August 18, 2011.
MAY, André *et al* . Effects of nitrogen and potassium doses and population density on the classification of onion bulbs. **HoRTic. BRAs.**, Brasília, v. 25, n. 3, Sept. 2007. Available at: http://www.scielo.br/scielo.php?script=sci_arttext&pid=S0102-05362007000300015&lng=en&nrm=iso. Accessed on May 18, 2011.
MEURER, E.J.; SBCS, Vigosa, 2006. **MiNERAL NuTRiZiOn OF PLANTS**, 432p. (ed. FERNANDES, M.S.
MONEGAT, C. **SOIL COVER PLANTS: CHARACTERISTICS AND MANAGEMENT IN SMALL FARMS**. Chapecó -SC. Ed. do Autor, 1991. 337p.
OLIVEIRA, Valter Rodrigues. **cuLTiVo DA cEBoLA.** Available at: http://www.unitins.br/ates/arquivos/Agricultura/Olericultura/Cebola/Cebola%20-%20Cultivo.pdf. Accessed on June 9, 2011.
RESENDE, Geraldo M de; COSTA, Nivaldo Duarte; PINTO, José Maria. Productivity and post-harvest quality of onions fertilized with increasing doses of nitrogen and potassium. **HoRTic. BRAs.**, Brasília, v. 26, n. 3, set. 2008. Available at <http://www.scielo.br/scielo.php?script=sci_arttext&pid=S0102-05362008000300019&lng=pt&nrm=iso>. Accessed on May 18, 2011.
RESENDE, L. M. de A; MASCARENHAS, M. H. T; SIMAO, M. L. R.. Overview of onion production and marketing in Minas Gerais. **iNFoRME AGRopEcuÁRio.** Belo Horizonte, v. 23, n.218, p.7-19, 2002.
ROSSIELO, R. O. P; JACOB NETTO, J. Aluminum toxicity in plants: new approaches to an old problem. **sBcs- NuTRizão MiNERAL DE pLANTAs** (Ed. Fernandes, M.S). Vigosa, MG: p.375-418, 2006.
SOARES, D.J., PITELLI, R.A., BRAZ, L.T., GRAVENA, R. and TOLEDO, R.E..B. **INTERFERENCE PERIODS OF WEEDS IN THE CROP OF TRANSPLANTED CEBOLA (ALLIUM CEPA L.).** Available at: http://www.scielo.br/pdf/pd/v21n3/a06v21n3.pdf. Accessed on September 1, 2011.
SOUZA, R. J. de; RESENDE, G.M. de. **CEBOLA CULTURE.** - Lavras: UFLA, 2002. 112p.
TOMÉ JUNIOR, J. B. **MANUAL FOR INTERPRETING SOIL ANALYSIS**. Guaíba : Agropecuária, 1997. ISBN 85-85347-19-8.
UFPR- Federal University of Paraná. **SOIL AMOSTRAGING: QUESTIONS AND ANSWERS** /. Department of Soils and Agricultural Engineering. Projeto de Extensao Universitária Solo Planta; (Coord.) Beatriz Monte Serrat .[*et al.*] -Curitiba : UFPR, 2002. 17p. :il. Available at: http://www.soloplan.agrarias.ufpr.br/amostragemsolo.pdf. Accessed August 3, 2011.
VIDIGAL, S. M.; PEREIRA, P. R. G; PACHECO, D. D. Mineral Nutrition and Fertilization of Onions. **INFoRME AGRopECUÁRIo.** Belo Horizonte, v. 23, n.218, p.36-50, 2002.
WORDELL FILHO, J.A,; BOFF,P. Diseases of parasitic origin. In: WORDELL FILHO, J.A.; ROWE, E.; GONQALVEZ, P.A.S.; DEBARBA, J.D.; BOFF, P.; THOMAZELLI, L.F. **MANEJo FITossANITÁRIo DA CULTURA DA CEBoLA.** Florianópolis: Epagri, 2006. 226p. p. 19-167.

# ANNEXES

**Annex 1:** Soil analysis report

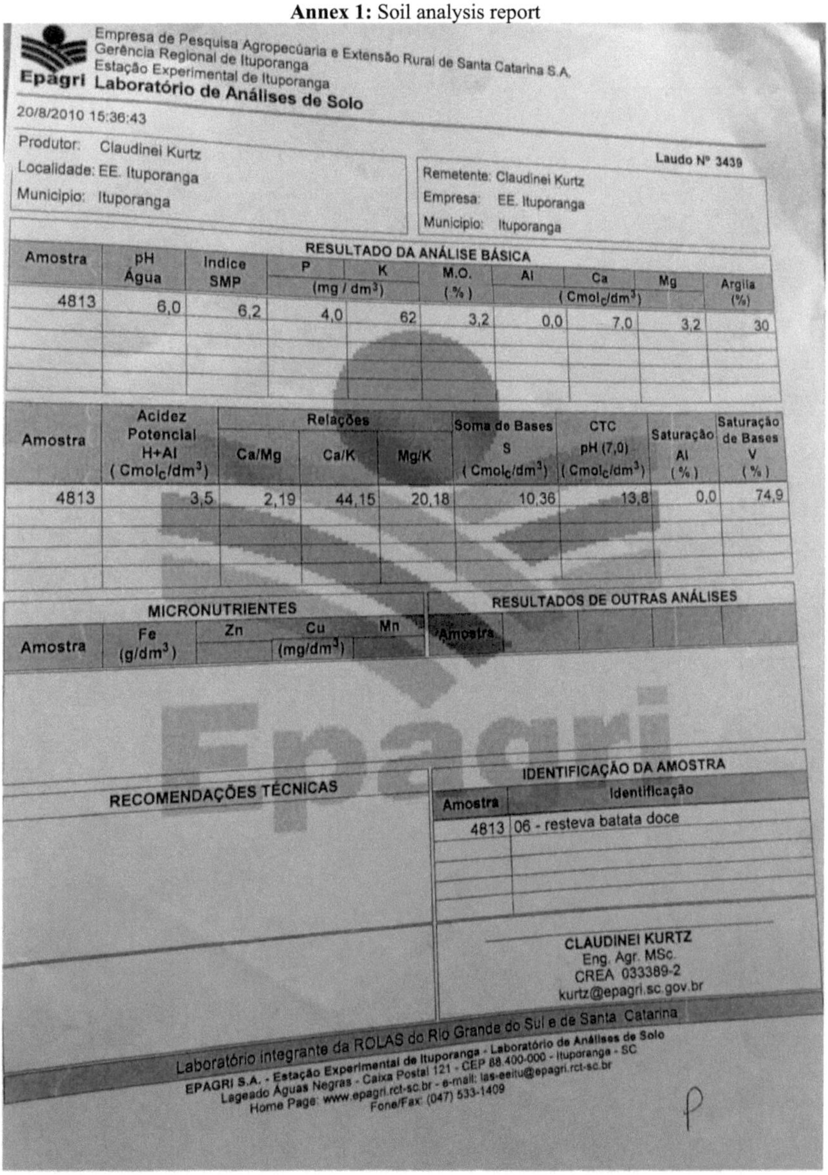

Empresa de Pesquisa Agropecuária e Extensão Rural de Santa Catarina S.A.
Gerência Regional de Ituporanga
Estação Experimental de Ituporanga
Epagri **Laboratório de Análises de Solo**

20/8/2010 15:36:43

Produtor: Claudinei Kurtz
Localidade: EE. Ituporanga
Município: Ituporanga

Laudo Nº 3439

Remetente: Claudinei Kurtz
Empresa: EE. Ituporanga
Município: Ituporanga

RESULTADO DA ANÁLISE BÁSICA

| Amostra | pH Água | Indice SMP | P (mg / dm³) | K (mg / dm³) | M.O. (%) | Al (Cmolc/dm³) | Ca (Cmolc/dm³) | Mg (Cmolc/dm³) | Argila (%) |
|---|---|---|---|---|---|---|---|---|---|
| 4813 | 6,0 | 6,2 | 4,0 | 62 | 3,2 | 0,0 | 7,0 | 3,2 | 30 |

| Amostra | Acidez Potencial H+Al (Cmolc/dm³) | Relações Ca/Mg | Relações Ca/K | Relações Mg/K | Soma de Bases S (Cmolc/dm³) | CTC pH (7,0) (Cmolc/dm³) | Saturação Al (%) | Saturação de Bases V (%) |
|---|---|---|---|---|---|---|---|---|
| 4813 | 3,5 | 2,19 | 44,15 | 20,18 | 10,36 | 13,8 | 0,0 | 74,9 |

MICRONUTRIENTES

| Amostra | Fe (g/dm³) | Zn (mg/dm³) | Cu (mg/dm³) | Mn (mg/dm³) |
|---|---|---|---|---|

RESULTADOS DE OUTRAS ANÁLISES

| Amostra | | | | |
|---|---|---|---|---|

RECOMENDAÇÕES TÉCNICAS

IDENTIFICAÇÃO DA AMOSTRA

| Amostra | Identificação |
|---|---|
| 4813 | 06 - resteva batata doce |

CLAUDINEI KURTZ
Eng. Agr. MSc.
CREA 033389-2
kurtz@epagri.sc.gov.br

Laboratório integrante da ROLAS do Rio Grande do Sul e de Santa Catarina

EPAGRI S.A. - Estação Experimental de Ituporanga - Laboratório de Análises de Solo
Lageado Águas Negras - Caixa Postal 121 - CEP 88.400-000 - Ituporanga - SC
Home Page: www.epagri.rct-sc.br - e-mail: las-eeitu@epagri.rct-sc.br
Fone/Fax: (047) 533-1409

Source: Kurtz (2011)

**Annex 2:** Field diary of activities carried out during the supervised curricular internship at EPAGRI - Experimental Station of Ituporanga - SC from 18/07/2011 to 05/12/2011, with a total of 360 hours.

| DATE | PEOPLE | CONTENTS | ACTIVITIES |
|---|---|---|---|
| 18/07/2011 | Staff | Experiment | Collection of soil from 0-10cm and 10-20cm in the experiment area; Demarcation of plots to implement treatments; Application of mineral fertilizer according to the design; |
| 19/07/2011 | | Office | Interpretation of soil analysis for onion crops following the recommendations of the liming and fertilization manual for the states of Santa Catarina and Rio Grande do Sul; Fertilization plan for onion crops; |
| 20/07/2011 | Agr.Eng. Édio Sgrott, Claudinei, Prof. IFC- Evandro | Office (center) | Visit to onion producers in the town of Bela Vista: Interpretation of soil analysis; Fertilization plan; Onion transplanting machine: mechanisms and forms of planting, performance in the field; |
| 21/07/2011 | | Course: Olericulture - cebóla | AEAVI onion course |
| 22/07/2011 | | Course: Olericulture - cebóla | AEAVI onion course |
| 25/07/2011 | | Office | Tabulated data from the phosphorus, potassium and manure experiments. |

| | | | |
|---|---|---|---|
| 26/07/2011 | Franscisco Gervini and Staff | | Implementing the experiment: fertigation X populated; marking out the plots; preparing the planting lines using a rotocaster attached to a microtractor; transplanting onions according to the design; |
| 27/07/2011 | Claudinei | Accompanying producer visits | Accompanying the visit of producers from Santo Amaro da Imperatriz in the state; characteristics of cover crops for use in SPDH and handling of machinery used in SPDH; |
| 28/07/2011 | | Office | Organized monitoring tables for the experiments; |
| 29/07/2011 | | Reading | Reading: Onion cultivation/ mineral nutrition; |
| 01/08/2011 | Employee | Laboratory | Weighing of crushed rye; Packaging in net bags; |
| 02/08/2011 | | Office | Organized folders for the experiments; Tabulated data |
| 03/08/2011 | Agr Eng: Gervini, Claudinei, Tec.Agritu Sementes | Visit to onion producers in Ituporanga | Arni Móhr- Cerro Negro and Laertes Móhr- Left Bank Rio Novo<br>Diagnosis: onion starts to develop slowly after transplanting, followed by death and stand loss; |

| | | | |
|---|---|---|---|
| 04/08/2011 | Claudinei Kurtz | Experiments | Inspecting the experiments, monitoring pests and diseases; Defining the schedule for applying fungicides, herbicides and fertilizers; |
| 05/08/2011 | | Office | Planning fertilization and weighing fertilizers to be applied in the UDESC experiment according to the design; |
| 08/08/2011 | | Office | Planning the implementation of an experiment on nitrogen fertilizer management; Reading reference material for the report; |
| 09/08/2011 | Flood | Flood | Flood |
| 10/08/2011 | | Office | Reading related articles |
| 11/08/2011 | | Office and farming | Typing and planning of the Nitrogen demonstration unit; Inspection of experimental areas and preparation of templates for application. Organic fertilization; |
| 12/05/2011 | | Office | Reading about onion crops and soil fertility; |

| | | | |
|---|---|---|---|
| 15/08/2011 | Employee | Farm and laboratory | Collecting nets with cover plants for drying in the greenhouse; Making templates for applying nitrogen in the fertigation/onion density experiment; |
| 16/08/11 | | Office | Reading and construction of the internship report; GPS survey of the experimental area; |
| 17/08/11 | | Office | Reading and writing an internship report; |
| 18/08/11 | | Laboratory / Office | Weighing material in nets to assess decomposition; Reading and assembling internship reports |
| 19/08/11 | | Soil collection | Soil collection for analysis; |
| 22/08/11 | | | Weighing nets Udesc experiment; reading, library |
| 23/08/11 | Claudinei Kurtz & Edio | Office/library | Reading, library; Interpreted soil analysis and fertilization plan for onion producers in Ituporanga; |

| | | | |
|---|---|---|---|
| 24/08/11 | Claudinei Kurtz | Office/laboratory | Making fertilizer templates; Visiting and monitoring experimental areas; |
| 25/08/11 | Employee | Fertilization | Fertilizer applied in the experiments according to the design and schedule; |
| 26/08/11 | Claudinei Kurtz | Farming | Visit to an onion producer's property in Atalanta-SC; onions in direct sowing showing nutritional deficiencies caused by the inordinate use of fertilizers; |
| 29/08/11 | Claudinei Kurtz | Experimental areas | Visit by extension workers from the Agroecology course at UFSC; Cover crops and soil management; |
| 30/08/11 | | Flood | Home activities, reading; |
| 31/08/11 | | Office | Reading and typing the report; |
| 01/09/11 | Claudinei Kurtz | Office/field | Monitoring experimental areas (application of gramicide and weeding of weeds not controlled by herbicides); Reading; |
| 02/09/11 | | Office | Reading and typing reports; |
| 05/09/11 | | Field/office | Collecting cover crops (rye); Monitoring the experimental areas; Tabulating data and graphs relating to the experiments; Fertirrigating the N experiment; |

| | | | |
|---|---|---|---|
| 06/09/11 | Claudinei Kurtz | Office | Discussion of data from the 2010 experiments; Data tabulation and reading; Monitoring the experimental areas |
| 07/09/11 | | National holiday | |
| 08/09/11 | | Flood | |
| 09/09/11 | | Flood | |
| 12/09/11 | | Flood | |
| 13/09/11 | | Flood | |
| 14/09/11 | | Flood | |
| 15/09/11 | Employee | Field / Laboratory | Collecting summer crop nets; inspecting experimental areas; weighing rye nets; various activities; |
| 16/09/11 | | Office | Tabulation of data relating to the experiments; monitoring of the experimental areas; |
| 19/09/11 | Staff | Experiment | Weighing of summer crops; application of N. experiment Density X N |

| | | | |
|---|---|---|---|
| 20/09/11 | Employee | Field and laboratory | Collecting rye straws for kiln drying; |
| 21/09/11 | Staff | Experiment | Fertilizer system installed in density x N experiment; |
| 22/09/11 | Staff | Experiment | Installed fertigation system experiment density x N |
| 23/09/11 | | Experiment | Nitrogen fertilization experiment N; Weighing of rye straws; |
| 26/09/11 | Staff | Experiment | Fertigation: application of N and K via drip irrigation; monitoring of experimental areas; |
| 27/09/11 | Staff | Office/experiment | Applied manure UFSC experiment; Reading and report; |
| 28/09/11 | | Office/experiment | Reading and report: monitoring experimental areas; |
| 29/09/11 | | Office/experiment | Reading and typing reports; |
| 30/09/11 | Employee | Experiment | Chlorophyll measurement in onion leaves, fertigation experiment; irrigated onion experiment N x Drip plant density; |

| | | | |
|---|---|---|---|
| 03/10/11 | Employee | Experiment | Fertilizing with N and K; N x Plant density experiment; Making labels to identify plots; |
| 04/10/11 | Employee | Experiment | Collecting rye; |
| 05/10/11 | Staff | Experiment | Irrigated Experiment N x Plant density; Calculation of flow rate and volume of water applied; |
| 06/10/11 | Staff | Experiment | Irrigated Experiments: P 1 and 2, K, Organo mineral; |

| | | | |
|---|---|---|---|
| 07/10/11 | Staff | Fertilization | Experiment N; Fertilizer dilution for fertigation; Monitoring experiments; |
| 10/10/11 | Employee | Fertigation | Application of N, K and Zn, B, and Mn via fertirrigation; Weighing rye paddocks; Organizing an experiment for a field day; |
| 11/10/11 | | Office | Reading and typing; |
| 12/10/11 | | Holiday | |
| 13/10/11 | | Office | Elaboration of labels to identify the experiments for the field day; |

| | | | |
|---|---|---|---|
| 14/10/11 | | Net collection | Collecting summer crops; |
| 17/10/11 | | Experiments | Identifying the experimental plots; attaching labels to the experimental plots; |
| 18/10/11 | | Experiments | Chlorophyll measurement experiment Fertigation; Identification of plots; |
| 19/10/11 | | Experiments | Collecting leaves for foliar analysis in P, K, Organomineral and fertigation experiments; |
| 20/10/11 | Claudinei, Paulinho And Gervini | Farm visit in Petrolandia-SC | Soil collection Physical and chemical analysis; Disease diagnosis (Sclerotium rolsfii). |
| 21/10/11 | Claudinei & Edio | Auditorium and experiments | Lecture to a group of Ituporanga onion producers on soil management and plant nutrition; visit to experimental areas; |
| 24/10/11 | Employee | Experiments | Fertigation experiment N X density; |
| 25/10/11 | Staff | UDESC experiment | Collecting, counting and identifying weeds and then drying and weighing them; |
| 26/10/11 | Employee | Laboratory | Grinding onion leaves for foliar analysis; |
| 27/10/11 | Employee | Experiment | Fertilizer application in the UDESC experiment; |

| | | | |
|---|---|---|---|
| 28/10/11 | | Experiment | Inspection of the experiments; |
| 31/10/11 | Employee | Experiment | Fertigation experiment N X density; |
| 01/11/11 | Employee | Experiment | Chlorophyll measurement exp. N x Density; |
| 02/11/11 | Holiday | | |
| 03/11/11 | Employee | Laboratory | Grinding onion leaves for foliar analysis; |
| 04/11/11 | | Experiment | Rye net collection;<br>Monitoring experimental areas; |
| 07/11/11 | | Experiment | Fertigation experiment N X density; |
| 08/11/11 | Claudinei/<br>Daniel/ Werner | Auditorium and experiments | Visit by students from the Agronomy / UFSC course, with a lecture and visit to the experiments; |
| 09/11/11 | | Experiment and office | Weighing rye straws: Typing and organizing the experiment tracking sheets; |
| 10/11/11 | | Office | Reading and tabulating data; |

| | | | |
|---|---|---|---|
| 11/11/11 | Employee | Experiment | Drip irrigation experiment N X plant density; |
| 14/11/11 | Employee | Experiment | Nutrient application via drip N X plant density experiment; |
| 16/11/11 | employees | experiment | Onion harvesting UFSC experiment; |
| 17/11/11 | | | Invitation to IFC Rio do Sul, evaluation of the horticulture technology course by MEC; |
| 18/11/11 | employees | experiment | Onion harvesting and evaluation of flowering plants P and K experiment; |
| 21/11/11 | | office | Reading and tabulating data; |
| 22/11/11 | | auditorium | Meeting of the onion sector chamber. Subjects discussed: - Evaluation of the harvest with analysis of harvest expectations; - lecture by Prof. Marcuzzo IFC, on *Sclerotium rolsfii* and other matters; |
| 23/11/11 | | Office/experiment | Evaluation of flowering plants; harvest planning |
| 24/11/11 | Employee<br>Claudinei and Jamil | experiment | Harvesting onions and evaluating flowering plants; UDESC and organomineral cover crops experiment; Visit to experimental areas by producers from the municipality of Angelina-SC; |

| | | | |
|---|---|---|---|
| 25/11/11 | employees | experiment | Harvesting onions and evaluating flowering plants N X plant density experiment in the fertigation system; |
| 28/11/11 | | office | Data tabulation |
| 29/11/11 | Staff | experiment | Bulb storage experiments: P,K and organominerals; |
| 30/11/11 | Staff | experiment | Evaluation and classification of bulbs from the experiments; |
| 01/12/11 | Staff | experiment | Evaluation and classification of bulbs from the experiments; |
| 02/12/11 | Staff | Office | Tabulation of data from P, K and organomineral experiments |

MIX
Papier aus verantwortungsvollen Quellen
Paper from responsible sources
FSC® C105338

Printed by Books on Demand GmbH, Norderstedt / Germany